Something New Under the Sun

Helen Gavaghan

Something New Under the Sun

Satellites and the Beginning of the Space Age

COPERNICUS
AN IMPRINT OF SPRINGER-VERLAG

Published in the United States by Copernicus, an imprint of
Springer-Verlag New York, Inc.

Copernicus
Springer-Verlag New York, Inc.
175 Fifth Avenue
New York, NY 10010
USA

Library of Congress Cataloging-in-Publication Data

Gavaghan, Helen.
    Something new under the sun : satellites and the beginning of the
space age / Helen Gavaghan.
        p.   cm.
    Includes index.
    ISBN 0-387-94914-3 (hardcover : alk. paper)
    1. Artificial satellites—History.   2. Astronautics and
civilization.   I. Title.
    TL796.G38   1997
    629.43'09—dc21   96-48689
                                                            CIP

Manufactured in the United States of America.
Printed on acid-free paper.
9 8 7 6 5 4 3 2 1

ISBN 0-387-94914-3   SPIN 10523953

Constance and James Gavaghan

Preface

When the story of our age comes to be told, we will be remembered as the first of all men to set their sign among the stars.

—Arthur C Clarke, *The Making of a Moon*, 1957

A new age was dawning, in which the organized brain power for military and civilian science and technology was the dearest national asset.

—Walter McDougall, ... *the Heavens and the Earth: a political history of the space age*, 1985.

When the Alfred P. Sloan Foundation decided to sponsor a series of histories of technology in the 1990s, they asked for proposals about technologies that have had a significant impact on the twentieth century. For some years, I had written news and features about space and had been hooked by the glamour of space exploration. Which aspect of the field would, I wondered, best fit into the Sloan's proposed series?

It seemed to me that the answer was navigation, weather, and communications satellites—that is, so-called application satellites. The National Academy of Engineering has said that of all the technological achievements of the second half of the twentieth century, these satellites are second only to the Apollo moon landing. Application satellites have a stealthy, silent influence on our lives. Most of us would notice them only in their absence. But then we *would* notice. There would be no early warning of hurricanes, no satellite data for the computer models that predict weather. There would be no instantaneous communication to and from any part of the globe, no satellite TV, and no navigation in bad weather. It would be a more dangerous and expensive world.

So I submitted a proposal. I wrote blithely of a history of every kind of civilian application satellite, from every country, from before the launch of *Sputnik* up to the 1990s. My book was also to encompass the critical supporting technologies of launch vehicles, electronics, and computers. Somehow, I won the grant.

After a few months of research, I discovered that I had known little about the subject and that it was full of apocryphal tales from imperfect memories. It took about three years to track down participants and locate archives, company records, and small pockets of papers kept by people when they retired. I had to find out what was classified and what wasn't, what people thought was classified even when it wasn't, and what, in general terms, might be in the genuinely classified material.

Not surprisingly, my original proposal was of absolutely no use. Its main fault was that it was about *civilian* application satellites. But navigation satellites were developed for a purely military purpose. The early history of weather satellites is inextricably intertwined with that of reconnaissance. And the decision that led to *Syncom,* the precursor of *Early Bird,* the world's first commercial communications satellite, owed much to the military's urgent need for improved global communication.

So the word *civilian* was the first thing to be excised from my conception of the book. It was followed by a ruthless culling of the 1990s, the 1980s, the 1970s, most of the 1960s, and satellites developed outside the United States. Finally, all but a few of the early American satellites fell by the wayside. Launch vehicles, electronics, and computers survived by the skin of their teeth, and only insofar as they demonstrated the limitations and difficulties surrounding those designing the early satellites.

What is left gives a flavor of yesterday's technology, which is our own technology in embryo, and a technology that has shaped our world. The book excludes many people, which is a shame but inevitable if it is to be readable.

The title, *Something New Under the Sun,* is a play on the biblical saying that there is no new thing under the sun. It was coined by Bob Dellar, an amateur astronomer who led a group of "Moonwatchers" in Virginia in 1956. The task of the Moonwatchers, who were scattered all over the world, was to track the satellites that the United States and the Soviet Union were planning to launch during the International Geophysical Year of 1957/58. Mr. Dellar is now dead, but Roger Harvey, who was sixteen at the time, was one of Mr. Dellar's group, and he mentioned the phrase in a parking lot in northern Virginia while we inspected the telescope he had used to search for *Sputnik.* I asked if I could purloin the phrase as the title of my book, and he said yes.

It is an apt title, because satellites were, literally and figuratively, something new under the sun. The pioneers who designed the first satellites admit cheerfully that they hadn't a clue what they were doing or what they were up against. Their launch vehicles blew up, their electronics were unreliable, guidance and control were primitive, the world was just turning from vacuum tubes to transistors, and those transistors didn't always work. The list of things they didn't know and that failed goes on and on and those things are, of course, the reasons why those early participants in the space age were pioneers.

The only non-American participant who is discussed at any length in the book is Sergei Korolev, the mastermind of the Soviet Union's space program who was responsible for the launch of *Sputnik*. He was an extraordinary man of extraordinary tenacity, who at great personal cost survived Stalin's paranoid and casual cruelties. Despite his contribution to the Soviet Union's Cold War armory, some tribute seemed called for, and the so-called chief designer of cosmic-rocket systems has the introductory chapter to himself.

Sputnik, according to the historian Walter McDougall, sparked the biggest furor in the United States since Pearl Harbor. The satellite was Korolev's baby, and it was launched as part of the International Geophysical Year.

The IGY was the brainchild of Lloyd Berkner, a leading American scientist. While scientists were still in the early stages of planning the IGY, President Eisenhower announced that the U.S. would launch scientific satellites as part of its contribution to the IGY. Within days the Soviet Union made a similar announcement.

It seemed at the time that the White House had bowed to pressure from industry, scientists, and the military. But recent scholarship suggests that President Eisenhower hijacked the IGY and made it, in the words of U.S. Air Force historian R. Carghill Hall, the stalking horse for the administration's plans for reconnaissance satellites.

As is so often the case, there were many people to whom it didn't matter at all why what happened, happened. The space age had opened, and the pioneers of navigation, weather, and communications satellites were ready.

At the Applied Physics Laboratory (APL) of the Johns Hopkins University, in Maryland, Bill Guier and George Weiffenbach listened to *Sputnik's* signal. Within the week, they were developing an approach to orbital determination that broke with a centuries-old tradition. Before Guier and Weiffenbach's work, the technique was to measure the angles to heavenly bodies and to determine orbits from those values. The scientists of the IGY had an elaborate optical and radio observational system in place for measuring the angles to satellites. Guier and Weiffenbach measured changes in frequency and developed computational and statistical techniques that at the time seemed to be coming from "left field."

Their boss, Frank McClure, adapted their techniques to form the basis for the Transit navigation satellites. In turn, Transit severed links with millennia of esoteric navigational rituals by providing mariners with computer readouts of latitude and longitude. Transit was developed because the Special Projects Office of the Navy needed some way to locate its Polaris nuclear submarines with greater accuracy than possible with existing methods. But the system went on to serve surface fleets, merchant vessels, the oil industry, fishing fleets, and international mapping agencies.

In the Midwest, Verner Suomi, of the University of Wisconsin, heard a lecture about the IGY and proposed flying an experiment to measure the radiation balance of the earth, a value of fundamental importance to meteorologists. The experiment set him on the path to earning the honorary title of father of weather satellites. These satellites took twenty years to find widespread acceptance among meteorologists. Because they rely on similar technology to that of reconnaissance satellites, they have a murkier history than that of satellite navigation.

In New Jersey, John Pierce (known in science fiction circles at the time as J.J. Coupling), of Bell Telephone Laboratories, played the pivotal role in the early days of the development of commercial communications satellites. He was swiftly challenged by Harold Rosen, of the Hughes Aircraft Company. On the title page of his book *How the World Was One*, Arthur C. Clarke calls Pierce and Rosen the "fathers of communication satellites."

Guier, Weiffenbach, McClure, Suomi, Pierce, and Rosen—these were the Edisons and Marconis of satellites for navigation, meteorology, and communication. Pierce and Rosen were rivals; Weiffenbach heard Pierce lecture and learned some things about satellite design; Suomi sought Rosen's help when he was trying to persuade NASA to fly another of his experiments; Weiffenbach met Suomi in India. They were not all close friends, but the American space community of the late 1950s was small and intimately connected. The same mysteries faced them all: the unimagined complexity of the earth's gravitational field, the unknown space environment and the radiation belts. All but Rosen benefited directly from the IGY. All were involved in projects that ultimately became the work of hundreds.

Something New Under the Sun is about the ideas of these men and the global, national, and local influences that shaped them.

In the case of Transit, most of the primary source material comes from APL and some of that material could be extracted for me only by people with the necessary security clearances, so there may well be things I am missing that I do not know about. The story is told through the eyes of APL, even though I have tried to set it in context. I'm sure someone viewing the story from outside APL would have a different tale to tell, but as written, I hope it gives a sense of what it was like to develop a satellite system in the late 1950s and early 1960s.

Meteorology satellites were more difficult to write about because the story is intricately linked with the change from art to science that meteorology was undergoing in the 1950s and because much of the primary source material was still classified when I was writing. But key participants helped to steer me through a sea of partial information. Verner Suomi is one of several who played a critical role in the early days, and perhaps more should be said about the others. If anyone writes the story at greater length, perhaps more will be said.

The early days of communications satellites are described mainly from the point of view of the Bell Telephone Laboratories and the Hughes Aircraft Company. Much of the text is based on material I collected from the archives and company records of AT&T and Hughes, supplemented by interviews and by other documents that participants passed on to me.

Echo and *Telstar* are names that still bring a flash of recognition to some faces. They are part of this book because John Pierce, whose ideas were important in many ways, was involved either directly or obliquely with them and because they highlighted AT&T's plans for global satellite communication, which raised antitrust concerns that shaped American policy in this strategically important new field. For these reasons, I have concentrated on *Echo* and *Telstar* rather than the NASA-sponsored Project Relay.

The most famous satellites that were based on Rosen's initial ideas were the *Syncom* satellites and *Early Bird*. These opened the era of commercial communications satellites.

Navigation, meteorology, and communication—ancients arts that have become sophisticated science and technology. The application satellites that ring the earth did much to help in that transition. Our social institutions and expectations are changing rapidly. Via satellite, we can remotely diagnose illness, watch from our living rooms while "smart" weapons shatter their targets, or track the development of major hurricanes for weeks before they threaten our coasts. The impact of satellites on our lives has scarcely begun.

Acknowledgments

Many people have helped me enormously: Richard Obermann, who patiently read many proposals and early drafts; Trish Hoard, who gave a friend's support; Nan Heneson, who edited sections and gave advice.

David Whalen provided much informed discussion on the subject of communication satellites and shared information he had gathered for his doctoral thesis from the NASA History Office and George Washington University.

Jim Harford allowed me to see an early draft of a chapter of his book about Sergei Korolev and shared an anecdote about Korolev that I have reproduced in Chapter 1.

Milton Rosen allowed me to interview him about his days on the Vanguard project, found old papers for me, and gave me a copy of his book on Viking and a copy of *Vanguard—A History*. Both were of great help.

Bill Pickering and John Townsend patiently explained their days with Vanguard and Explorer and took my phone calls when I was stuck on some detail.

Roger Harvey, an amateur astronomer and "Moonwatcher," gave me the inspiration for the title of the book. He, Henry Fliegel, and Florence Hazeltine responded to an advertisement for Moonwatchers and told me of their experiences as members of the program.

For the navigation section, a particular acknowledgment of Bill Guier and George Weiffenbach is gladly given.

Thanks to Bill Guier for meeting me at APL and for the hours of interviews and for patiently and repeatedly explaining details of physics.

Thanks to George Weiffenbach and his wife for traveling to Providence, Rhode Island, to meet me for a weekend and for the hours of interviews, dinners, and lunch discussions; for the participation in phone calls too numerous to mention; and for reading and checking the passages concerning physics.

Thanks to members of the Transit team for their time. Bob Danchik, Bill Guier, George Weiffenbach, Lee Pryor, Carl Bostrom, Henry Elliott, Lee Dubois, Charles Pollow, Laurence Rueger, Tom Stansill, Russ Bauer, Charles Bitterli, Harold Black, Ben Elder, Eugene Kylie, Edward Marshall, George Martin, Barry Oakes, Charles Own, Henry Riblet, Ed Westerfield.

All of the above except Bill Guier, George Weiffenbach, Tom Stansill, Eugene Kylie, Charles Bitterli, and Henry Riblet spent a day in a conference room with me at the Applied Physics Laboratory near Baltimore, Maryland.

What they said, which I recorded on several hours of tape, is scattered in numerous small details throughout the chapters of the navigation section. The day was invaluable because it gave a wonderful sense of the camaraderie and creative tensions that existed between them and of the intensity with which they must have worked on Transit.

Their faces and voices when they spoke of Richard Kershner, the team leader, and of Frank McClure provided the information—if something as fleeting and intangible as facial expressions can be called information—on which I based comments about McClure and Kershner. Both died at comparatively young ages, and their technical writings give little away about their personalities.

Of the people at the roundtable, Bob Danchik and Lee Pryor subsequently extracted unclassified reports from amongst classified documents in the APL archives. They also responded to my many phone calls as I struggled with the intricacies of the subject.

Thanks to Carl Bostrom for drawing my attention to the poem that completes the navigation section.

Thanks also to Henry Elliot for locating the papers he had saved from his days with the Transit program and for the interview at his home in Washington, DC.

Thanks to Eugene Kylie for traveling into Washington to discuss the Transit receivers over lunch.

Tom Stansill generously gave telephone interviews and mailed me old brochures and papers about the early Transit receivers.

Thanks to Charles Bitterli for the interview and lunch in Silver Springs, Maryland, and for anecdotes about the coding of the early algorithms.

John O'Keefe provided reminiscences about uncovering the pear-shaped nature of Earth, the presatellite days, and his impressions of Bill Guier and George Weiffenbach. And thanks to his wife for the ride back to the metro and stories of her research into Emily Dickinson's life and work.

Gary Weir, a historian at the Naval Historical Center, in Washington DC, discretely steered me in the right direction on the subject of Polaris.

Acknowledgments

Thanks to Commander William Craft, of the United States Naval Academy, in Annapolis, for some general discussions about navigation and the use to which the U.S. Navy put Transit.

Thanks also to Brad Parkinson, who headed the development of the Global Positioning Satellite.

Group Captain David Broughton, director of the Royal Institute of Navigation, pointed me in the direction of British officers who were once connected with Special Projects and who still feel the need for discretion.

Phil Alberts, the archivist at the Applied Physics Laboratory, helped me in working with the APL's archives, and Elaine Frazier assisted me in extracting information from the archives.

The meteorology section would have been impossible without the hours of interviews with Verner Suomi.

Dave Johnson was invaluable in helping me to put the early days of meteorology satellites into context. He and Pierre Morrel helped to explain some basic ideas of meteorology, and if their efforts failed, it is my fault.

Thanks also to Tery Gregory for diving into the basements of the University of Wisconsin in search of Professor Suomi's files.

Thanks to R. Cargill Hall for drawing my attention to his paper on the interrelationships between the IGY and President Eisenhower's national security policy.

John Pierce allowed me to interview him, found old papers and copies of books, and gave me a ride back into Palo Alto. Thanks for all of these.

At the Hughes Aircraft Company, Pat Sinclair ferreted through company records and provided leads. Bob Roney handed over copies of papers he had cherished since his days on the Syncom project. Harold Rosen explained something of his participation in the Syncom project and tried to make communications satellites less of a mystery to me.

A special thanks to John Rubel and his wife, who gave me a room for a week in their delightful home outside Santa Fe, which allowed me to rifle through John Rubel's extensive records of his days as deputy director of defense research and engineering. These were documents which he has had declassified since leaving the Pentagon.

Finally, thanks to Bill Frucht and Sally Gouverneur, my editor and agent.

Contents

3   Communications   169

What has been is what will be, and what has been done is what will be done; and there is nothing new under the Sun.

—Ecclesiastes 1:9

Prologue:
The Seed

The scientific mind is a curious thing. It probes what others take for granted, including, on one night in 1950, a multilayered chocolate cake. Some of America's brightest scientific minds were focused on that cake at the home of James Van Allen, who was to become famous as the discoverer of the earth's radiation belts and who was hosting a dinner for the eminent British geophysicist Sydney Chapman. With admirable attention to detail, the collective scientific intellect verified that the cake had twenty-one layers. That cake did much to put the scientists in the kind of mood from which expansive conversation flows and big ideas are born.

Led by Lloyd Berkner, the talk turned to science. Berkner was a charismatic individual, head of the Brookhaven National Laboratory and a veteran of one of Admiral Byrd's Antarctic expeditions. Like his fellow diners, Berkner was fascinated by the new insights into the earth's environment that instruments aboard V2 rockets captured from Germany had been providing since 1946. These high-altitude sounding rockets were invigorating the earth sciences. Was the time right, asked Berkner, to promote an international effort in geophysics, one that would exploit new and established technologies in a world-wide scientific exploration to illuminate global physical phenomena?

This conversation was to lead to the establishment of the International Geophysical Year of 1957/58. The IGY was the enterprise that by inadvertently dovetailing with a key element of President Eisenhower's national security policy—establishing the freedom of space for reconnaissance satellites—was to become the cradle of the space age.[1]

---

1. The Eisenhower administration wanted to establish the freedom of space by launching a civilian scientific satellite. Such a satellite would pave the way for America to launch reconnaissance satellites that could fly over foreign territory without eliciting international protest or retaliation. Only a few of the scientists participating in the IGY knew of the administration's secret purpose, and it is not yet clear who these were. In fact, it is only now that historians are beginning to fully uncover the political relationship between the IGY and the reconnaissance satellite program. See . . . *the Heavens and the Earth: a political history of the space age,* by Walter A. McDougall (Basic Books, 1985), which contains an extensive review of Eisenhower's intelligence needs and posits, on the basis of documents then

At Van Allen's dinner party in 1950, Berkner and his fellow diners were not considering spacecraft, though they all knew of the imminent technical feasibility of launching satellites and were to play important roles in the opening years of the space age.

Berkner's suggestion for a geophysical year captured Chapman's attention. These two agreed to "talk the idea up" among their many contacts in the international scientific community. Chapman sent an account of the dinner party to the journal *Nature*. Within a few years, Berkner and Chapman had secured enough interest to win the backing of the International Council of Scientific Unions for the IGY. Chapman became president of the IGY and Berkner the vice president.

Scientists had already established the idea of international collaboration in 1882 and 1932 when they had undertaken to study geophysics from the earth's poles. Expanding the concept of the International Polar Years to encompass the whole earth was, agreed the diners, a good idea, and the best time for such an effort would be between July 1, 1957, and the end of 1958. They chose the dates to coincide with a period of maximum solar activity, when there would be much to study.

During solar maxima, which occur once every eleven years, the sun throws out huge flares of matter and energy more frequently than at other times, adding peaks of intensity to the solar wind that is always racing through the solar system. Numerous terrestrial effects result. The northern lights, for example, move to lower latitudes than usual as more charged particles precipitate along magnetic field lines into the ionosphere, an area of charged particles at altitudes of between 60 and 1000 kilometers above the earth's surface. Studying the ionosphere was to account for a significant portion of the IGY, including science that was important for long-range communication, missile development, and, as it turned out, the space age.

Others in 1950 wanted to set up observation posts in places where meteorological data were sparse. Some wanted to organize expeditions to places where they could observe total eclipses. And some, like Lloyd

available to McDougall, that the Eisenhower administration wished to finesse a satellite into orbit in order to establish the freedom of space. See also *The Eisenhower Administration and the Cold War, Framing American Astronautics to Serve National Security,* by R. Cargill Hall (in *Prologue, Quarterly of the National Archives,* spring 1996). Cargill Hall's article, which is based on additional, recently declassified documents, makes the argument more explicitly that the IGY and national security policy were linked.

Berkner, were intrigued by the physics and chemistry of the upper atmosphere.

So, with endorsements from the international scientific community, the participating countries got down to business. Each needed a detailed research plan. And they needed money.

In the U.S., scientists turned for leadership to Joseph Kaplan, a professor of physics at the University of California, Los Angeles. He had proven his organizational abilities by cofounding the university's Institute of Geophysics in 1944 and by campaigning successfully for degree programs in sports. As chair of the U.S. National Committee for the IGY, he concealed the habitual tension that turned him into a five-cigar man during sporting events.

The first meeting of the U.S. National Committee for the IGY took place at the National Academy of Sciences on the afternoon of March 26, 1953. For a day and a half, the group struggled to define a program and budget. Frustration mounted. One scientist suggested that they should forget the whole thing. Berkner, a consummate committee man, smoothed over such moments, and by the end of the twenty-seventh they had outlined their aims. In the meantime, they had solicited thoughts from their colleagues around the country.

Ideas poured into the academy during April. Many of them would sound familiar today. Issues were raised that today's scientists continue to address. Paul Siple wrote, "There is evidence that the earth is undergoing a significant climate change, advancing from cooler to warmer conditions ... our knowledge is still imperfect as to the exact cause of climate changes."

By May the list of subjects to be studied included geomagnetism, solid-earth investigations, atmospheric electricity, climatic change, geodetics, cosmic rays, ionospheric physics, high-altitude physics, and auroral physics.

It is hard to imagine any study of these subjects today that would not rely partially or wholly on satellite observations. Then satellites did not exist, and the National Committee of the IGY did not initially consider that satellites should be developed. It was, however, clear to them that they needed a strong program of sounding rockets to probe the upper atmosphere. Some scientists were concerned that the rockets would cost too much, and perhaps price dissuaded them from adding satellites to their agenda.

The estimated price tag for this unprecedented research proposal, between 1954 and its closure at the end of 1958, was $13 million—$2.5 million more than the National Science Foundation (NSF) was requesting as its total budget for fiscal year 1954. The NSF and the academy undertook the task of requesting the money from Congress.

Behind the scenes, Kaplan lobbied hard. There were times when the project seemed doomed. A scientist in the administration told him, "Joe, go home. Such a beautiful program does not stand a snowball's chance in hell of getting support." Yet, it did.

Through the end of 1953 and 1954, planning for the IGY went ahead in the U.S. Concurrently, support for launching a satellite, despite deep skepticism from many, was gaining momentum among a vocal and persuasive minority in the military, industry, and academia. Some saw the IGY as the natural home for a satellite program, and they set in motion events that led to the General Assembly of the IGY endorsing the inclusion of satellites in its program.

This was in Rome in the fall of 1954. The night before, Berkner and others who were to become leading space scientists had debated until the early hours whether their enthusiasm was getting ahead of their ability, and whether they should seek the approval of their international colleagues. Slowly the enthusiasts converted the cautious. As they argued and won their case the next day, the Soviets simply observed—in silence. It was October 4, three years to the day before the launch of *Sputnik I*.

The International Geophysical Year was now close to its decisive encounter with the Eisenhower administration's national security policy. While Berkner and his colleagues prepared for the Rome meeting, a top-level scientific panel, authorized by President Eisenhower in July 1954, was in the process of assessing the technological options available to prevent a surprise attack on the U.S., particularly by the Soviet Myacheslav-4 intercontinental BISON bombers. This panel, headed by James Killian, a confidant of President Eisenhower and the president of the Massachusetts Institute of Technology, separated its task into three areas: continental defense, striking power, and intelligence capabilities.

Their final report contained a recommendation that the administration fund development of a scientific Earth satellite to establish the freedom of space in international law and the right of overflight.

The report was finally completed in February 1955, and later that month, Donald Quarles, the assistant secretary of defense for research and development, discreetly asked some members of the U.S. National Committee for the IGY to formally request a scientific satellite.[2]

In response to the Rome resolution, the national committee had already asked a rocketry panel, which met for the first time on January 22, 1955, to " ... study and report on the technical feasibility of the construction of an extended rocket, from here on called the long-playing rocket, to be launched in connection with scientific activities during the International Geophysical Year." "Long-playing rocket" was a euphemism for satellite launcher, and the new panel was told that its discussions should remain private.

By early March the panel had concluded that a satellite launch was feasible within the time frame of the IGY, but that guidance would be difficult. On March 9, the executive committee of the U.S. national committee debated whether to accept the panel's findings. After some heated discussion, the report was accepted and the national committee requested that a satellite program be included in the IGY.

Quarles took the request to the National Security Council, which accepted the IGY's satellite project on May 26, 1955. The next day, Eisenhower, too, endorsed the project. On July 29, President Eisenhower told the world that the U.S. would launch a satellite as part of the IGY. Thanks to Eisenhower's national security aims, the scientists who had campaigned for science satellites had what they wanted. Many of them, of course, did not know of the underlying agenda that had furthered their aims.

Within days of Eisenhower's announcement, the Soviet Union, by then a participant in the IGY, talked openly about its satellite plans at a meeting of the International Astronautical Federation in Copenhagen.

Leonid Sedov, an academician and chair of the impressively named Commission for the Coordination of Interplanetary Flight, made the announcement. He dropped into one of the sessions and, through his interpreter, called a press conference at the Soviet embassy. During the conference, Sedov said that the Soviet Union planned to launch a satellite in about eighteen months, six months earlier than the earliest American

2. The information about the Killian panel's recommendations and the approach that Quarles made to members of the U.S. National Committee of the IGY comes from R. Cargill Hall's article in the *Quarterly of the National Archives.*

estimate. The plans he outlined were for a much larger satellite than those the U.S. was planning. For those whose job it was to analyze Soviet intentions, here was public confirmation of the heavy-lift launch capabilities that the Soviets aspired to and a clear announcement that missiles capable of intercontinental distances were in the offing.

Sedov's prediction of the Soviet launch date was wrong by ten months. But the Soviets did launch first, and their satellite was bigger than anything American scientists really believed in their hearts that the Soviets were capable of.

New Moon

If they had known then what they know now, would they have done the job? They would surely have been daunted, even given the imperatives of the Cold War. But they did not know how difficult space exploration would be. The "cold warriors" had their incentives: intercontinental ballistic missiles and reconnaissance satellites. And the enthusiasts, who sometimes were also cold warriors, had their long-held aim: to go beyond Earth's atmosphere. By Thursday evening, October 3, 1957, their destination was less than a day away.

That same evening was one of the last on which delegates gathered at a conference organized by the National Academy of Sciences in Washington D.C. as part of the International Geophysical Year. The focus of the conference was on the rockets and satellites to be launched before the end of 1958. The IGY united sixty-seven nations in a seemingly impressive *pax academica,* yet, despite good intentions, the satellites of the IGY were about to push the world into a new phase of the Cold War.

Satellites were not part of the original plan. Indeed, when the IGY first endorsed satellites and chose as its logo a satellite orbiting the earth, many considered satellites to be little more than science fiction. In three years that had all changed. By October 1957, the U.S. and USSR were following one another's progress keenly. Each group of scientists wanted to be the first, and the meeting at the National Academy of Sciences gave them all an opportunity to probe each other's intentions.

The Soviets were not very specific. Their delegation, led by Lieutenant General Anatoly Blagonravov, knew that a launch was imminent and had announced this on the first day of the conference. But they had not said exactly when. It is doubtful that they knew.

The Soviets would be gratified if the launch took place during the meeting, but they also knew, as the defector George Tokady was to say years later, that " . . . the launch of *Sputnik* was too big a piece of cake to play games with." Already the centennial celebrations on the birth of Konstantin Tsiolkovsky, Russia's "father of spaceflight," had passed on September 17th without a satellite attaining orbit. The cognoscenti had speculated that the seventeenth might be the occasion for a Soviet launch.

But the satellite would be launched when all was ready, and one man would make that determination. Late Thursday afternoon, as the conference workshop on rocketry struggled with the usual committee minutiae, that man, Sergei Pavlovich Korolev, lay—perhaps—restlessly in bed. More likely, being who and what he was, he paced the sitting room of his small cottage. It would be nearly a decade before the Soviet leadership publicly acknowledged the existence of this man, whom they called the Chief Designer of Cosmic-Rocket Systems.

Korolev's cottage was on the grounds of Baikonur Cosmodrome, near the village of Tyuratam in Kazakhstan, one hundred miles east of the Aral Sea. It lay on a parallel with northern Wisconsin and as far east of the Greenwich meridian as Halifax, Nova Scotia, is west.

In Washington D.C., as October 3 drew to a close, workshop coordinators pulled ideas together for the next day.

For Korolev, it was the early hours of October 4. That day Korolev's team would open the space age.

In future years, on the nights before he sent cosmonauts into space, Korolev would sleep very little. Instead, he worried about the well-being of the young men whom he drove as hard as he drove himself, buoyed by an energy that made his days so much longer than those of other people. But his wakefulness had a second purpose. It helped him avoid dreams in which guards beat him and screamed at him, dreams from which he would wake to a visceral fear of annihilation. These were Stalin's legacy, as were his memories of colleagues who one moment were working alongside him and the next were gone.

On October 4, as the sun rose on a clear, cool dawn, Korolev was—comparatively—free. On his wall hung a portrait of Tsiolkovsky. Tsiolkovsky had died twenty-two years before, but he had been the first to write scientifically about this day. His books, calculations, and ideas had long ago seduced Korolev.

Less than a kilometer away, beyond the small knot of trees that surrounded his house, the rocket glinted in sunlight. Beyond that, where only two years before no launch site had been, the semiarid steppes of Kazakhstan stretched to the horizon and beyond. The American government had known of this launch site, built with forced labor, since the spring of 1957, when a U2 spy plane had returned with photographs.

For Korolev there must have been tension, anticipation, excitement, and fear as he faced the day that would test all that he had become and all that he had dreamed.

As Korolev prepared to leave for the launch site, perhaps he thought briefly of the people waiting for him to fail, the naysayers and his rivals. Earlier in the year, as rocket after rocket had exploded and Nikita Khrushchev's impatience had mounted, Korolev's detractors had pushed for his dismissal. Then, in early August, his team had successfully launched the R7, the world's first intercontinental ballistic missile. Not for eighteen months would the U.S. launch a missile capable of similar range. A second successful launch followed, and on August 27 the Soviet Union announced that it possessed intercontinental ballistic missiles. It would have been more accurate to say that they had the beginnings of the capability, but the success had satisfied Khrushchev, who now had a rocket that would eventually be capable of carrying a two-ton thermonuclear warhead to the heart of America. As a result, Korolev had won the final go-ahead for his dream, the opportunity to send an artificial moon into space.

The following two months were full of frantic activity. Korolev's team had immediately begun intense preparations for a launch, and toward the end of August the satellite was ready to ship to the Cosmodrome. Korolev moved to his cottage to supervise launch operations. In September, the pace picked up and tempers frayed. The satellite developed an electrical fault. Everyone panicked. Korolev, always relentlessly demanding, became merciless. Time and again, the launch team watched with apprehension as his little finger rose to stroke his eyebrow; it was the signal to move smartly to the next job. They knew his capacity for compassion and for explosions of wrath; that he found people who would argue a case interesting, but would take personal offense at anyone who did not do his job. Korolev drove his engineers and his engineers drove themselves until, crisis by crisis, they coaxed the novel technology to readiness.

At last, toward the end of September, the crane in the assembly building hoisted the small, shiny sphere into the nose cone of the rocket. The launch team was now ready to put on a show for VIPs from Moscow. Members of the State Commission, the secret group that was to control the country's space program, flew in. Technicians ran a final check on the satellite's radio transmitter, switching the signal to a loudspeaker so that the commissioners could hear it echo around the building. Then the engineers silenced the transmitter. Some recalled later that their skin tingled at the thought that the satellite would not speak again until it was in space.

On the night of October 2, the launch team moved the rocket. It was four stories high and weighed nearly three hundred tons. Slowly, so very slowly, they wheeled the rocket out of the assembly building on a flatcar, and it began its painstakingly careful journey down the railroad track to the launch pad. It swayed with each uneasy movement. The next morning, Thursday October 3, they began the final preparations for launch—the countdown.

As the day progressed, the launch team worried that the satellite would overheat despite the gaseous nitrogen circulating inside the sphere. They threw a white blanket over the nose to give the satellite further protection from the sunlight. Later, dissatisfied with that solution, they pumped compressed air around the nose cone.

On Korolev's recommendation, the satellite ensconced in the nose cone was of a very simple design—a two-foot diameter sphere weighing 184 pounds. A sphere, Korolev said, was a fitting shape for what might be the world's first satellite because it mimicked the shape of the natural bodies of the universe. Consummate engineer that he was, it seems that he could never quite suppress the poet in himself.

Soviet scientists planned both optical and radio tracking for their satellite, with the intention of learning what they could about the earth's gravitational field and the density of the upper atmosphere. That week in

Washington, Soviet delegates were reemphasizing the frequencies that their first satellites would transmit. And the conference workshop on tracking resolved to establish stations capable of tracking the Soviet satellites, a resolve that, even as they made it, was too late.

The Soviets called their satellite *Prosteyshiy Sputnik* (meaning "simplest satellite"). Inevitably, the design and launch team had shortened that to "PS"; and Korolev's staff, who referred to him informally as "SP" (for Sergei Pavlovich), used the initials interchangeably for the man and the satellite.

PS's surface was buffed, as were those of the American satellites, so that it would shine in orbit as a sixth-magnitude star and could be tracked visually. It had four whiplike antennas that were pressed between the inside of the nose cone and the satellite's surface. When, or if, *Prosteyshiy Sputnik* attained orbit, these antennas would relay radio signals at a frequency that every amateur radio operator in the world would be able to detect.

Prosteyshiy Sputnik's destination was not far away, for the boundary of space is less distant from Earth than New York is from Washington D.C. Yet every inch of that journey would be fought against Earth's overwhelming gravity, which would yield only reluctantly to human ingenuity. If the launch vehicle did not reach a high enough velocity, PS's trajectory would be a ballistic path high through the atmosphere and back to the Earth's surface, like that of the ICBM that Korolev had launched in August. Alternatively, PS might be released into too low an orbit and burn up in the atmosphere.

Prosteyshiy Sputnik's launcher was designed to put a far heavier cargo in space, but Korolev was moving conservatively. The heart of the launcher was the R7 ICBM; four cone-shaped, strap-on boosters surrounded its base, resembling a stiff pleated skirt. The boosters would peel away when their fuel was spent, leaving the R7, minus the burden of the boosters' weight, to make the final push to orbital velocity. At liftoff, the R7 and the boosters, each containing a cluster of engines, would push from the Earth with more than half a million kilograms of thrust: power enough, if the launch was successful, to boost Nikita Khrushchev's domestic reputation and help him to ward off those remaining critics who had participated in a failed attempt to oust him from power just a few months earlier. Khrushchev's improved status should, in turn, help Korolev win backing for a continuing space program. A failure would set back Korolev's aspirations, for space was not Khrushchev's dream.

As it turned out, the wave of excitement that was to sweep the globe when *Sputnik* was launched seemed to take Khrushchev—and President Eisenhower—unawares. *Prosteyshiy Sputnik,* the simplest satellite, changed completely the world public's perception of the Soviet Union's technical capabilities, and Khrushchev learned quickly that space could yield political advantages. As a result, Korolev's aspirations were to be harnessed tightly to Khrushchev's international political goals, often in ways that cost lives and held back scientific advances.

On the morning of October 4, Korolev knew none of this. He needed success. But Korolev was also a dreamer, on a grand and generous scale, and he had dreamed this dream for thirty years. He had first worked with rockets after graduation from the Moscow Higher Technical School. He had stood on a sidewalk in Moscow as Friedreich Tsander, a fellow dreamer and space pioneer, raised his fist and said, "Forward to Mars!" That was in 1930, after the two young men had spent the evening with friends planning a research group to develop rockets and rocket-assisted aircraft. They called themselves the Group for the Study of Reaction Propulsion (GIRD).

Korolev regarded Tsander, who talked of rockets as though they already existed, as an older brother. At first, the two men and their friends had worked with no official backing or financial support in the cellar of an abandoned warehouse. They soon attracted the attention of the Soviet armaments minister, Mikhail Tukhachevskiy, a stroke of good fortune that won the group financial success but within a few years would lead to tragedy. In the meantime, GIRD expanded and joined the Gas Dynamics Laboratory in Leningrad to form the Rocket Research Institute. Korolev was appointed the deputy director, responsible to Tukhachevskiy.

Before GIRD moved to Leningrad, they designed, built, and tested the Soviet Union's first liquid-fuelled rocket. Korolev lit the fuse. The rocket, which was based on many of Tsander's ideas, flew successfully on August 17, 1933, and landed 164 yards from the launch site.[1] Sadly Tsander, who had died in March at the age of 46, did not witness the short flight.

1. Robert Goddard launched the world's first liquid-fuelled rocket on March 16, 1926. It reached an altitude of 41 feet and landed 184 feet from the launch site.

Despite what must have been grief at the death of his friend, petty restrictions limiting access to foreign journals, and the scarcity of food, even with ration cards, those were good years for Korolev. The government funded the institute because rocket research conformed to the national political goal of establishing Soviet technical supremacy. In Leningrad, Korolev met Valentin Glushko, another Soviet space pioneer who designed rocket engines. In future years, Chief Designer Korolev was to collaborate often with Glushko, though the two men were to develop a stormy relationship. At the same time, Korolev's personal life expanded. He married his school sweetheart, Xenia Vincentini, a surgeon, and they had a daughter, whom they named Natalia. But when Natalia was three, everything changed.

In the early hours of June 27, 1938, Stalin's secret police, the NKVD, arrested Korolev. Though he did not know it at that moment, this was the end of his marriage and the beginning of torture, hunger, and years of imprisonment. He was charged with anti-Soviet activity, and his guilt was determined apparently by his association with Mikhail Tukhachevsky. Tukhachevsky was already dead, condemned and shot a year earlier on the strength of false documents that some suspect had been planted by the Nazis.

The NKVD packed Korolev into a boxcar of the Trans-Siberian Railway destined for Magadan. From there he was transported in the hold of a prison ship to the gold mines of Kolyma, concentration camps where thousands died each month. For nearly a year he was hungry, far hungrier than in the ration-book days of the early thirties. He lost his teeth, developed scurvy, and in winter often woke to find his clothes frozen to the floor; but he survived.

He survived because the authorities transferred him to Moscow, to another kind of prison, one that held the cream of the Soviet Union's aeronautical designers. Andrei Tupolev, the country's most eminent aircraft designer at that time, headed the technical work of these scientists and engineers, though he was himself a prisoner.

When new prisoners arrived from the Gulag, Tupolev would ask them for a list of the engineers whom they had left behind in the camps. Some were reluctant to make such a list in case their colleagues had been freed and would be rearrested. That was probably not Tupolev's intention, and he may have seen Korolev's name on such a list and asked for him to be transferred to Moscow. Tupolev would have recognized the name because he had supervised Korolev's diploma project—the design of a two-seater glider—during Korolev's final year at the Moscow Higher Technical School.

Whatever the reason for the transfer, Korolev found himself working long hours in a Moscow prison with sparse comforts. The authorities had a twofold work incentive scheme. They held out the hope of eventual freedom, and they threatened the prisoner's families. Compared to Kolyma, it was paradise. But now Korolev knew that life could change any moment at the whim of a faceless bureaucrat. He tried repeatedly to impress this knowledge on fellow prisoners—that they might disappear without trace and no one would know of it. That knowledge was the darkness that would follow Korolev through his life, the darkness he would share with his friends in future years when late-night conversation lasted into the early morning hours.

Korolev spent his days like the other prisoners, designing aircraft for the war effort; at night, in the communal dormitory, he worked on rocketry. With the end of World War II, his rocketry research once more emerged into daylight. The USSR and the U.S. were vying for men and equipment from Peenemünde, the base where Germany had developed the rockets that bombarded London, Paris, and Antwerp in a new kind of warfare. As the Red Army swept into eastern Germany, Stalin remembered the rocketeers he had imprisoned eight years earlier. Back then a magistrate had told Korolev, "We don't need your fireworks and firecrackers. They are for destroying our leader, are they not?" By 1945, Stalin had changed his mind, and he turned to those rocketeers who had survived his purges.

Stalin was far more strongly committed to missile development than was the U.S. leadership at that time. In 1945, Stalin insisted on seeing prisoner Korolev. Korolev always remembered how without taking his pipe from his mouth Stalin had demanded information about the potential speed of missiles, their range, payload, and accuracy. In his memoirs Khrushchev says that the rationale for official interest was that the U.S. could, if it wanted, station long-range bombers at air bases in Europe close to the Russian border, whereas the Soviet Air Force could not reach the continental U.S.

Korolev was sent under guard to Germany to glean what he could. He was under orders to track down those V2s that were not on their way to America and to select German engineers from among those who had not surrendered to American troops. Korolev sent both rockets and engineers to Russia. In the years immediately after World War II, both the Soviet Union and the U.S. learned as much as they could from German advances in rocketry while simultaneously developing their own missiles. In Russia, the Germans worked separately and did not know what Soviet engineers were doing, which disappointed western intelligence workers when Stalin sent the Germans back to Germany in 1951. Yet this should not have surprised them. Von Braun, who worked on intermediate range ballistic missiles in the U.S., could likewise not have known details of the Americans' work on ICBMs.

By 1951 Korolev was no longer a political prisoner, at least not obviously so. He and Xenia had divorced in 1946, and he had since remarried. He worked fanatically hard, prompting colleagues to wonder whether he had a home life. Of prison he rarely spoke. Occasionally, he would sip cognac late at night and reminisce with other former prisoners, telling them how those days still haunted him in his dreams. And a few nights before he died,[2] he told two close friends, the cosmonauts Yuri Gagarin and Alexei Leonov, of some of the pain. His death on an operating table in 1966 was, some speculate, a result of poor health stemming from his days in Kolyma.

In many ways Korolev's early life must have prepared him for privation.[3] He was born in Zhitomir, in Ukraine, in 1906. Three years later his parents separated, and Sergei went to live with his maternal grandparents while his mother went back to college. She left instructions that the child was not to leave the garden to play because she was afraid his father, who had already threatened her with a pistol, might kidnap him.

Korolev remembered those years as lonely ones. For a few years he wrote poetry. But he remembered vividly that when he was six, his grandmother who was fascinated by gadgets and technology took him to see his first aircraft. This experience touched Korolev as deeply as seeing ballet or theater might define the destiny of another human being.

On weekends his mother would visit. Then they would play, and she would read to him, but she also made him go alone to distant dark rooms

---

2. Leonov told this story to Jim Harford, who is writing a biography of Sergei Korolev to be published by John Wiley and Sons in October 1997.

3. Much of the information about Korolev's early life is drawn from Yaroslav Golvanov's book—*Sergei Korolev: The Apprenticeship of a Space Pioneer.*

because, inspired by her reading of James Fenimore Cooper, she thought that in this way he would conquer fear.

As they did to millions of others, World War I and the Russian Revolution turned Korolev's life upside down. In 1917, Sergei was eleven. By then his mother was married again, this time to an engineer called Grigory Balanin. The family lived in Odessa, on the Black Sea, and Grigory and Sergei settled to the prickly business of getting to know one another. The world war and foreign occupying forces depleted food stocks. Civil war followed world war. The family was often hungry, and Korolev would hike with his mother to the countryside to barter for potatoes.

When the civil war ended, with Lenin in charge, Sergei's mother and Balanin sent Korolev to the First Construction School in Odessa. Here he learned physics and mathematics (one of the school's mottos was "mathematics is the key to everything") and how to tile roofs. He soon decided that he did not want to be a roofer, because by now he was fascinated by the idea of flying and designing aircraft and gliders.

After an obligatory summer as a mediocre roofer, Korolev applied to and was accepted by the Kiev Technical School. Two years later, he moved to the Moscow Higher Technical School where he earned his diploma in aeronautical engineering. A year later Korolev graduated from flying school. In the meantime he earned money by working in an aircraft design bureau.

He was now twenty-four, interested in rockets, but passionate about designing and flying sophisticated, record-breaking gliders. And in this role, anyone who cared to watch would have seen the character of the future chief designer emerge. Oleg Antonov, another of Russia's great aircraft designers and also a gliding fanatic, met Korolev at about this time. Korolev was attempting a record flight in a glider of his own design. Inadvertently, Antonov sent Korolev aloft with the anchor still attached to his glider. Korolev flew for four hours nineteen minutes, oblivious of the anchor. When he landed and saw the hole in the tail of his glider, he offered to tear Antonov's eyes out with a pair of pliers. Yet Antonov remembered Korolev as a man of iron will and boundless humor.

By the morning of October 4, 1957, that combination of will and humor had carried Korolev to the position of chief designer of rocket-cosmic systems. Stalin had been dead for four years. After Stalin's death, Korolev had been invited to join the Communist Party and had been elected a corresponding member of the USSR's Academy of Sciences.

Ostensibly, he was a secure member of the establishment, yet the injustice of wrongful imprisonment ate at Korolev. He had asked repeatedly, and without success, to be rehabilitated—for his conviction at Stalin's hands to be rescinded. Only after the successful launch of the satellite now sitting in the nose of the rocket on the launch pad would this happen.

So, on the morning of October 4, 1957, it was both the chief designer and the prisoner who awakened in terror, who turned a collar to the cold, climbed into his car, and drove to the concrete apron of the launch site. To the colleagues waiting for him, he was SP, a man rapidly becoming a legend among the few who knew of him. He was short and heavyset. He reminded them of a boxer or a wrestler, as much because of his personality as because of his physique. His brown eyes were bright with intelligence and passion.

Korolev saw himself as the designer whose role was to define the job, to listen to the team members, and then to make the decision. This he had done, paying attention to strategy and detail, balancing the consequence of one technical choice against another, seeking the right compromise. Now the product of hundreds of people's work, of thousands of calculations, of designs and redesigns, was sitting on the launch pad. Would it open a new frontier?

Perhaps for propaganda reasons, perhaps because there would be an emotional symmetry in such an event, and perhaps because it is true, there is a persistent and disputed story that while at the Moscow Higher Technical School, Korolev had visited Tsiolokovsky. The old man, so the story goes, told Korolev that rockets were a very difficult business, and Korolev had replied that he was not afraid of difficulties. It is hard to imagine that en route to the launch complex, Korlev did not remember Tsiolkovsky, who, when he died in 1935, was an old and nationally revered man. Perhaps, too, Korolev remembered Friedreich Tsander, his old friend from the free days in Moscow.

On October 4, 1957, Tsiokovsky's, Tsander's, and Korolev's dream stood against the gantry, gleaming in the sunlight.

In Washington D.C., delegates awoke to the penultimate day of the rocket and satellite conference. A workshop was to debate what to include in the IGY's manual on rockets and satellites.

Korolev drove to the launch pad. The countdown was proceeding, and they were about to fuel the rocket. He mounted the platform to brief the engineers and to listen to accounts of the night's doings. Then, while the launch team pumped liquid oxygen and kerosene into the tanks, he called Moscow with an updated report of the countdown.

Throughout the day, Korolev monitored everyone's work, outwardly calm but fooling no one. By evening, technical difficulties had halted the countdown several times. Those who could stayed out of Korolev's way.

The day moved inexorably forward. A day that for Korolev must have lain before him as a path to forever then passed in a second.

Thirty minutes before liftoff, everyone retired to their posts. Most went into a concrete bunker one kilometer from the launch pad. Those without a part to play in the final countdown climbed onto the bunker's roof.

Korolev sat at a desk in the bunker, watching through a periscope. Floodlights bathed the rocket. Hope was palpable. As they waited and watched, someone walked beneath the floodlights. The unknown figure raised a bugle and blew clear notes into the midnight sky before hurrying back to safety.

Korolev listened as the loudspeakers relayed the deliberately spoken script of a space launch, a script that is still running, but which played that night to its first audience.[4]

"Duty crew, leave the pad."

"Fire brigades, on alert."

"Zero minus one minute."

"Switch to start vents."

Korolev knew that nitrogen was sweeping through the pipes, purging the giant rocket before oxidant and fuel met in a mighty chemical reaction.

"Auxiliary engines pressurized."

"Main engines pressurized."

"Start."

---

4. The words in quotation marks are extracted from Golovanov's lengthier account, which appears in *Sergei Korolev: The Apprenticeship of a Space Pioneer.*

Stillness enveloped the watchers. They dared not blink. And then it happened. Incandescent vapors engulfed the rocket, throwing stark shadows on the surrounding concrete. The earth rumbled and a thunderous roar washed past their ears. They watched the huge rocket strain, and then—as if in slow motion—the engines lifted the rocket from the earth.

Did Korolev remember what Tsiolkovsky had written? "Mankind will not remain on the earth forever, but in the pursuit of light and space, we will, timidly at first, overcome the limits of the atmosphere and then conquer all the area around the sun." Well it had begun. And what would those earlier versions of Korolev have thought of his fifty-one-year-old self, sitting, eyes glued to a periscope, watching his dream and an ancient dream of humanity's ascend? Would the frozen wretch in Kolyma, with hunger griping in his belly, have believed this moment? What of the man who would start awake in terror or the child who saw with joy his first aircraft? What of the teenager in a civil war, the student, or the test pilot? Each gave a gift to the chief designer; surely they watched with wonder what together they had wrought?

Korolev slowly came to himself. Only minutes had passed, and already the rocket was a distant point of light. Around him people hugged and kissed, unshaven chins scraping cheeks a little damp. They danced and shouted, "Our baby's off!" Soon they fell silent and listened. The loudspeaker reported all systems nominal, later that the rocket had reached orbital velocity, and then that the satellite had separated from its rocket.

Now they faced another wait. Was the satellite in orbit? It should be overhead again in about ninety minutes. As the time approached that the satellite should be coming into range, they looked gravely at the radio operator. Then they heard the distinctive beep of their satellite in orbit, the signal they had last heard in the assembly building on that day a lifetime ago. The earth had a new moon. *Sputnik I* was in orbit.

Korolev notified Khrushchev and received the first secretary's muted congratulations. The party machinery swung into operation. Soon the world's teleprinters would carry news of the triumph and Soviet propaganda. Editors around the world would be galvanized as they read reports that included the words, "Artificial Earth satellites will prove the way for space travel, and it seems that the present generation will witness how the freed and conscious labor of the people of the new socialist society turns even the most daring of Man's dreams into reality."

At the cosmodrome, Korolev returned to his engineers. The chief designer, who had already experienced tragedy, now knew triumph. He was to know both again. That night, he mounted a platform and thanked his staff, those present and those at home. He was radiant. He continued, "Today we have witnessed the realization of a dream nurtured by some of the finest men who ever lived, including our own Konstantin Eduardovich Tsiolkovsky. Tsiolkovsky foretold that mankind would not forever remain on the earth. The sputnik is the first confirmation of his prophesy. The conquering of space has begun. We can be proud that it was begun by our country. A hearty Russian thanks to all."

Cocktails and the Blues

2

"I think there is very little doubt that the Russians intend to
start firing [satellites] sometime on or before the first of the
year [1958]."

—Richard Porter, chairman of the U.S. Technical Panel on Earth Satellites,
October 3, 1957

"As it happened, the public outcry after *Sputnik* was earsplit-
ting. No event since Pearl Harbor set off such repercussions in
public life."

—William McDougall, . . . *The Heavens and the Earth, a Political History
of the Space Age*

Lieutenant General Anatoly Blagonravov sipped vodka. In Tyuratam it
was the early hours of Saturday, October 5, 1957. But in Washington,
D.C., it was still the evening of Friday, October 4, and Blagonravov was
hosting a reception at the Soviet embassy for delegates to the IGY's conference on rockets and satellites. *Sputnik I* was in orbit, had, in fact, already
passed undetected over America. It would not again go unnoticed.

The people at the Soviet embassy did not yet know that the space
age had begun. As the guests circled the buffet of Russian delicacies spread
beneath sparkling chandeliers, the Americans probed Blagonravov for
hints of when his country planned to launch a satellite.

Although Blagonravov headed the Soviet delegation, the Americans
could not have known quite how ideally placed he was to answer them.
He was an academician and chair of the Soviet Academy's Interdepart-
mental Commission of Interplanetary Communication; as such he knew
what the Soviet scientists wanted to do with satellites. He was a former
head of the rocketry and radar division of the Soviet Academy of Artillery
Sciences, to which Sergei Korolev had been elected in 1954. Finally,
Blagonravov was a member of the State Commission and one of the VIPs
who had listened to Prosteyshiy Satellite's test signal in the assembly building at Tyuratam little more than a week earlier.

21

What no one who saw Blagonravov that night can say is whether he knew at the beginning of the reception that *Sputnik* was already aloft. The white-haired, scholarly figure mingled genially with his guests, his demeanor much the same as it had been all week. His blue eyes masked an inner intensity. His Russian cigarette was tilted between his lips.

What is indisputable is that Blagonravov knew that a launch was imminent. So did the American scientists—at least intellectually—including William Pickering, a native New Zealander of great charm and underlying sharpness. In 1957, Pickering was a veteran rocket scientist and had been the director of the Jet Propulsion Laboratory in Pasadena, California, for three years. He worked closely with the German rocket pioneer Wernher von Braun. These two knew that with comparative ease they could convert one of the intermediate range ballistic missiles (range 2,000 miles) they were developing into a satellite launcher.

Pickering had first become involved with rocketry in 1942, at the Jet Propulsion Laboratory (JPL). The lab had analyzed intelligence reports about a V2 rocket that had landed in Sweden. In 1943, Army Ordnance asked JPL to investigate the possibility of developing an American long-range rocket, which in those days meant rockets that could fly at least one hundred miles.

When the war ended, von Braun surrendered to American troops and was moved eventually to the Army Ballistic Missile Agency in Huntsville, Alabama. The U.S. Army also shipped home about one hundred V2s, which von Braun had designed. These rockets were distributed to groups around the U.S. that were interested in upper-atmosphere research and rocketry. A group of scientists known as the Upper Atmosphere Research Panel selected and analyzed the experiments that the rockets would carry. Members of this group, including James Van Allen, went on to head the IGY's satellite efforts.

Pickering, in cooperation with the Army Ballistic Missile Agency and von Braun, was soon testing the V2s while JPL—known as the "army smarts"—simultaneously continued its own rocket development. At times, as rocket after rocket exploded, it seemed to Pickering that rocketry was an unpredictable art, not a science. But JPL and the Army persevered.

By the early 1950s, the V2s were proving the truth of what Sergei Korolev had said in 1934 in his book *Rocket Flight in the Stratosphere:* " . . . a rocket is defense and science."

Von Braun, too, had always known this. Like Korolev, he dreamed of space and worked on missiles. In his new country, von Braun was a highly visible advocate of space exploration. In the early 1950s, he drew headlines by promoting grandiose ideas of space stations and Martian colonies at a time when the general public and most scientists thought that even a simple satellite was science fiction.

While the idea of a space station was indeed science fiction at that time, the technical pieces that would make the launch of a satellite a practical project were in place in 1954. That September, von Braun and a small group of scientists drew up a plan for a satellite that would become famous in the annals of space history. His idea was that a Redstone rocket would lift a five-pound satellite from Earth and that an upper stage, comprising a cluster of Loki rockets atop the Redstone, would give the satellite its final kick into orbit. If the small satellite was successful, a second, larger, "bird" carrying scientific instruments would follow. This proposal was called Project Orbiter.

The result of the first shot would be five pounds of metal circling hundreds of miles above the earth at a speed of roughly 17,000 miles per hour, more than four miles per second. Impressive, but how could they prove that such a small, swift, and distant object was really in orbit?

Nowadays, of course, radio links satellites to observers on the ground. Then radio was not an obvious choice. Von Braun turned to astronomy. The satellite would, after all, be a new heavenly body.

No one was better qualified for the job of finding a body speeding through space than the man he turned to, Fred Whipple. Whipple was the director of the Smithsonian Astrophysical Observatory in Cambridge, Massachusetts. A tall, slender man with a widow's peak and steel-rimmed glasses that gave him an air of erudition, he was expert at tracking meteors and comets. Whipple concluded that optical tracking could follow a burnished satellite at an altitude of two hundred miles, even though the satellite would be visible only for the twenty minutes at dawn and dusk that the sun would illuminate it against a dark sky. So the proposal for Project Orbiter included no plans for radio tracking, a decision that—seemingly—was to tell against von Braun less than a year later.

Von Braun sent the proposal to the Department of Defense, JPL, and branches of the armed services. Whipple approached the National Science Foundation and the National Academy of Sciences.

The proposal came at a time when others with influence were strongly advocating that the U.S. should begin a satellite program. Among these were the scientists who had gained the support of the international scientific community for the inclusion of a satellite in the IGY, and, secretly, those in the Air Force who wanted satellites for reconnaissance. In addition, the Naval Research Laboratory (NRL), in Washington D.C., made its own pitch for a satellite program and launcher.

The NRL had been one of the groups to receive V2s, and under the technical leadership of Milton Rosen, it had developed its own rocket—the Viking—for upper-atmosphere research. Rosen persuaded a few forward-looking thinkers to write about the job that satellites could do when rockets were developed that could reach orbit. Among these was John Pierce, from AT&T's Bell Laboratories, who, along with Harold Rosen from the Hughes Aircraft Company, would later be christened "the father of communication satellites" by the science fiction writer Arthur C. Clarke.

It was the IGY's scientists who won Eisenhower's public support, and on July 29, 1955, the president announced that the U.S. would launch a satellite as part of the IGY. He deferred a decision as to which of the armed services would provide the launch vehicle.

The day before the formal announcement, the White House press office summoned journalists, telling them that the briefing was embargoed for the next day. The day's grace was to allow American scientists time to notify their colleagues overseas. One scientist flew to New York to give a letter for Sydney Chapman, president of the IGY, to a friend who was flying to London.

In a two-hour background briefing at the White House, James Hagerty, Eisenhower's press secretary, announced that America planned to launch ten basketball-sized satellites sometime during the International Geophysical Year. The journalists rushed to the phones, only to be brought up short by the locked doors. Hagerty was determined that the journalists should understand that the story was not to be broken until after the official announcement the next day. One journalist asked what he, a crime reporter, was supposed to do with the story.

Jet Propulsion, the sedate journal of the American Rocket Society, stopped the presses. The editor wrote, "We cannot imagine anything more exciting than to look up and see through our binoculars the brilliant point of light in the sky that will represent a new astronomical body created by man. We expect to be deeply moved."

Until the announcement, only about a hundred people knew of the satellite plans. Now the secret was out. But the launch vehicle decision was still pending.

The unenviable task of recommending a launch vehicle fell to a panel of scientists headed by Homer Joe Stewart. Stewart, widely referred to as Homer Joe, was well respected for his technical expertise. He was an aerodynamicist from the California Institute of Technology and a prime mover on the Upper Atmosphere Research Panel. Homer Joe's panel was guided in its deliberations by two precepts: satellites must not interfere with missile development, and the project should have a strong civilian flavor and scientific component. Both charges to the Stewart committee must surely have stemmed from the more secret deliberations within the White House that the Killian report stimulated.

The first requirement was largely taken care of when the Air Force, which had earlier been assigned responsibility for military space projects, dropped out of the competition, saying that it could not develop a rocket in time to launch a satellite during the International Geophysical Year without compromising its missile development. Not a surprising decision, given that the development of intercontinental ballistic missiles had top priority with the Department of Defense and that parts of the Air Force were lobbying for funding to develop reconnaissance satellites.

The second requirement was more difficult to fulfill. Only the military were then sponsoring launcher development.

But the Army and the Navy squared off: the ring, the committee rooms of Washington D.C. and private offices in the Pentagon.

In one corner was Wernher von Braun with an updated version of Project Orbiter. At his back, the formidable figure of General John Medaris and the technical expertise of the Jet Propulsion Laboratory. The laboratory had suggested important changes to the upper stages of the launcher, but the size of the proposed satellite—five pounds—stayed the same.

Nevertheless, the Army's proposal had an immediate strike against it, because the Army was developing intermediate-range ballistic missiles. While not as high a priority as those that the Air Force was developing, they were undoubtedly missiles and compromised perception of Orbiter as a civilian project.

In the opposing corner was the Naval Research Laboratory with its proposal, called Project Vanguard. That proposal contained a much stronger scientific component than the Army's, a component that the Naval Research Laboratory, with its strong background in upper atmosphere research, was well qualified to implement.

The laboratory had additional advantages. It had successfully developed the Viking sounding rocket as a replacement for the V2. The Viking was not part of a weapons system but was intended for scientific exploration of the upper atmosphere—albeit in support of military purposes.

Further, the Viking project had provided the lab with a record of developing a large project on time and to cost. The Glenn L. Martin Company had been the contractor for Viking, and the laboratory proposed to work with them again on Vanguard.

Finally, the NRL had what was for that time a sophisticated radio tracking plan, whereas the initial Project Orbiter included only optical tracking.

Much of Project Vanguard was technically innovative, relying on miniaturization to beef up the satellite's capabilities. While miniaturization is today a technological cliché, in 1955 the concept was new. Vanguard would have tiny batteries and radio equipment and in principle would be able to carry ten pounds of scientific instruments, far more than Project Orbiter offered. The batteries were chemical because the designers decided that solar-battery technology, generating electricity from sunlight, was unlikely to be practical in the time available.

The Navy won the first round, but the Army bounced back with a plan that took account of technical criticisms, including an improved tracking proposal. Milton Rosen, who had headed the Viking development and would be the technical director of Project Vanguard, listened to them argue their case for most of an afternoon. He watched tensely, afraid that von Braun's eloquence was swaying the panel.

In the event, Homer Joe's committee confirmed its original decision, and on September 9, the secretaries for the Navy, Army, and Air Force were notified that the Navy would head the three services in the development of a satellite launcher. When the news reached the NRL's site by the Potomac in Washington D.C., there was jubilation, some surprise, and a conviction that they could do the job.

The decision of Homer Joe's panel was not unanimous. Homer Joe himself was one of two who thought that the Army should get the job. The nation could save time and money, he thought, by piggybacking the satellite program on the Army's missile work. Many, then and now, believe that he was right, and that if the U.S. had picked Project Orbiter, America would have been the first into space.

Thus Project Orbiter became famous as the proposal that could have begun the space age but did not. In retrospect, and in the light of the emerging evidence concerning the Eisenhower administration's desire to establish the freedom of space by launching a civilian scientific satellite, one suspects that the Army could not have won. Certainly, the guidance given to the Stewart committee seems to have been fashioned to favor the somewhat less militaristic Navy proposal.

The administration's decision to back Project Vanguard has pulled down much vilification on President Eisenhower's head. Though it seems now that Eisenhower's decision was more subtle than it appeared at the time, he did underestimate the blow to America's self-esteem and the concomitant gain in the Soviet Union's international prestige that would follow the *Sputnik* launch.

During 1955, as wheels of policy turned within wheels, the scientists who in the fall of 1954 had won backing from the international scientific community for the inclusion of satellites in the IGY had also been busy. Early in 1955, the executive of the national committee had established a rocketry panel to evaluate the feasibility of producing a rocket that could launch a satellite. The task was delegated to William Pickering, Milton Rosen, and a young hawk, also from the Naval Research Laboratory, John Townsend.

They met in Pasadena in February of 1955. This trio concluded—unsurprisingly—that a satellite launch was feasible. Milton Rosen reported their findings to the rocketry panel in March 1955. He outlined three possible rocket configurations, which, unbeknownst to many scientists of the IGY, had much in common with proposals which were then vying for attention at the Pentagon.

On March 9, Joseph Kaplan, chair of the U.S. IGY, sought approval from the U.S. executive committee for the IGY for the inclusion of satellites in their research plans.

He met opposition. The debate, recorded in scribbled, handwritten notes, is hard to decipher, but the language suggests that some of the scientists knew or had an inkling that national security issues were at stake. Merle Tuve, director of the Department for Terrestrial Magnetism at the Carnegie Institution (and founder of the Applied Physics Laboratory, which would develop the Transit navigation satellites), was very uncomfortable with the idea of tying the satellite project to the IGY. Others, like Fred Whipple and Harry Wexler, who as chief scientist of the weather bureau would be a strong supporter of meteorology satellites, argued persuasively that satellites would expand the science that could be undertaken by the IGY.

His colleagues asked Tuve—wouldn't it be better for satellites to be part of the IGY so that the data collected by the spacecraft would be unclassified and freely available? If the Department of Defense were to lead the way, there would be less opportunity for international collaboration. They could do an enormous amount of science with satellites, they argued. And if they were to exclude everything from the IGY that had a possible military application, there would be nothing left. Who were they, asked one scientist, to hold up progress?

Hugh Odishaw, the indefatigable secretary of the national committee, finally summarized their options: say no, irrespective of whether the satellite project had scientific merit or not; embrace the satellite project as part of the American contribution to the IGY; decide whether it had merit and, if it was geophysically useless, say that the satellite program was more appropriate to the Department of Defense.

Clearly, satellites do have considerable potential for geophysics, and what seems to have persuaded Tuve to endorse the satellite project was that if the Pentagon took the initiative, cooperation with other countries would be difficult. Eventually, the executive committee approved the inclusion of satellites in the IGY.

The National Science Foundation then requested funds from Congress on behalf of the National Academy of Sciences, which was organizing the IGY (the National Academy of Sciences does not request funds directly from Congress). The National Science Foundation and the National Academy of Sciences gave Donald Quarles, assistant secretary of defense for research and development, what he needed, two bodies with indisputable credentials in the world of civilian science. He now had a formal request for a civilian satellite as had been recommended by the Killian committee.

Kaplan sent a letter to the National Science Foundation explaining that the satellite project would need $10 million in addition to the nearly $20 million that congress had by now authorized for the IGY. At the same time, the IGY's satellite proposal was sent to the White House and the Pentagon. At the Pentagon, Quarles looked at Kaplan's figures and doubled them. Even that turned out to be a serious underestimation of the ultimate cost (elevenfold what Kaplan requested, excluding the Pentagon's expenditure), inevitably, perhaps, because no one had any idea of the technical difficulties and costs of space exploration.

It was now May 1955, and the IGY's organizers needed the money as soon as possible if they were going to launch a satellite before the completion of the IGY at the end of 1958. But the project stalled. At least, that is how it appeared to the staff of the U.S. national committee's secretariat, who were frustrated as phone calls went unanswered and unsatisfying memos arrived in response to requests for information about the satellite budget.

In fact the IGY was now a pawn in a bigger game. The chain of events was in progress that led ultimately to the National Security Council's endorsement of the plan at the end of May, to the formation of Homer Joe Stewart's committee, to President Eisenhower's announcement of July 29, and to the September decision that Project Vanguard would launch America's first satellite.

From the IGY's perspective, things started to move again after Eisenhower made his announcement. Despite not knowing which launch vehicle the administration would select, the national committee asked Richard Porter, a technical consultant to General Electric, to form a panel to oversee the IGY's satellite work. Porter's committee spawned two more. One, led by William Pickering, was responsible for radio and optical tracking and orbital computations. Fred Whipple was one of his committee members, and he took on the responsibility of organizing optical tracking for the IGY. James Van Allen chaired the second group, which was responsible for choosing the experiments that would fly.

As 1956 advanced, the cost of the satellite program escalated. Eventually, one of the IGY scientists complained that the satellite program was costing as much as if each satellite had been built from solid gold. Berkner assured him that the value of the satellites to the international community and to the United States was worth far more than digging that much gold out of the earth.

Despite such enthusiasm, development of the Vanguard rocket was experiencing technical and financial difficulties. Milton Rosen, Vanguard's technical director, was convinced that the Glenn L. Martin Company was not giving the NRL its best efforts. He attributed this to the company's also having the contract for the Air Force's Titan ICBM. Whatever the reasons, the pitfalls and colossal nature of undertaking to develop and build a three-stage rocket from scratch in so short a time were now apparent.

These technical difficulties were compounded by the cutbacks in the program. By October 1957, Van Allen's committee could count on only six launch vehicles, possibly far fewer.

Sputnik was about to change all that.

On Friday, October 4, 1957, Walter Sullivan, a science correspondent for the *New York Times,* filed a story for Saturday's paper. After a week at the rockets and satellite conference he felt fairly certain that he had taken the right line. He'd written that the Soviets were close to a launch attempt.

Sullivan says that before going to the reception he called his office. He was told that the paper's Moscow office had heard Radio Moscow announce that a satellite was in orbit. He was told to find out what he could.

Sullivan hurried to the embassy. He found the American scientists: Pickering, Richard Porter, Lloyd Berkner, and John Townsend. When Sullivan told them the news, they looked at one another in consternation. To this day, Pickering believes that the Soviets at the reception did not know of their own successful launch.

The Americans went into a huddle and decided that they must congratulate their hosts. As the senior American scientist present, the task fell to Lloyd Berkner. Leaving his colleagues to assimilate the news as best they could, he took his glass and a spoon and climbed on a chair. He rapped the glass for attention. Slowly the conversation died down, and the guests turned to Berkner. His announcement was simple, and it caused a sensation. He said, "I am informed by the *New York Times* that a satellite has been launched and is in orbit at an altitude of nine hundred kilometers. I wish to congratulate our Soviet colleagues on their achievement." With that, he raised his glass. Blagonravov was beaming as he drank the toast.

Follow That Moon

I think we very likely face the embarrassing situation that, say, next spring, we have one or two [satellite tracking] cameras and the Russians have one or two satellites. . . .

—William Pickering, October 3, 1957

Pickering drank the toast. He felt curiously detached from his surroundings. What thoughts did the shock loosen? Probably something like, " . . . it could have been us, what now, they said imminent, they told us. . . . " Certainly he knew better than anyone else present that his laboratory, working with von Braun's team, might already have had a satellite aloft. And he had speculated often enough with colleagues at the Jet Propulsion Laboratory that the Russians must be ready for a launch soon. They'd feared the event, but they hadn't truly believed it could happen.

Pickering remained deep in thought. When he looked around, his colleagues had gone. He knew where. Pickering followed to the IGY's offices a few blocks from the Soviet embassy in downtown Washington D.C. There they sat, Dick Porter, Homer Newell, John Townsend, and Lloyd Berkner. Three years earlier, some of them had been together in a hotel room in Rome, where they had plotted late into the night how they would win backing from the General Assembly of the International Geophysical year for a satellite launch. All had an emotional investment in the unfolding events. All had campaigned to persuade their government to build a satellite. And when President Eisenhower gave the go-ahead, they'd fought amid the turbulent waters of internecine and interservice rivalry. They had not wanted the same launch vehicle and satellite, but in the end they had found common cause. Together, they faced a bitter moment.

They looked at one another and asked, Now what? Into this introspection the telephone blared. NBC had interrupted radio and television programs with the news. At the American Museum of Natural History, in New York, the phone rang every minute. Most calls were from people wanting to know how to tune into the satellite's signal or when they could see it. For the first night at least, curiosity was uppermost, though some called to say that the stories couldn't be true, the Russians could not have beaten the U.S. As yet there was no panic or concern; that came the following week.

At the United States' headquarters of the IGY, the little group remembered its priorities, decided early in their planning of Vanguard. First, place an object in orbit and prove by observation that it was there. Second, obtain an orbital track. The satellite's path would give valuable knowledge about the earth's gravitational field and the density of the upper atmosphere. Finally, perform experiments with instruments in the satellite. Here, too, orbital tracking and prediction would be important. An instrument isn't much good if you don't know where it is when it records a measurement.

All of this was for the future. That Friday night, the first priority had to be met. It wasn't an American satellite, but something was aloft. Was it in a stable orbit, or would it plummet to Earth within hours? They let the task of learning as much as possible about the satellite's orbit chase bitterness and disappointment away for a short time. The task was difficult, because the launch had caught them unawares.

And what was the task? They wanted to find and follow an object of indeterminate size, traveling several hundred miles above the earth at about 17,000 miles per hour. Ideally, they would have liked first to pinpoint *Sputnik*'s position (to acquire the satellite) then to observe parts of its orbit (to track the satellite) and to determine from those observations the parameters of that orbit.

So that they would know when and where to look to acquire the satellite, they needed to know the latitude and longitude of the Soviet launch site, and the time, altitude, and velocity at which the satellite was injected into orbit.

Imagine an analogous situation. Hijackers have taken control of an Amtrak train and no one knows where or when this happened, nor which track the hijackers have coerced the driver to follow. How does Amtrak find its train? The company could alert the public to look for its train or, if the train had a distinctive whistle, the public could listen for it. When the alert public had called in the times and places at which they spotted or heard the train, Amtrak could calculate roughly where the train would be, given an estimate of speed and a knowledge of the network of tracks. The greater the accuracy with which observers recorded the time and place of the train's passing, the more accurately Amtrak could predict its train's future location.

With an exact location of the launch site and information about the position and time of the satellite's injection into orbit, the group at the IGY could have predicted *Sputnik's* position, then tracked its radio signal and determined an orbit. But the Soviets did not release this information. In fact, they did not publicly give the latitude and longitude of the launch site for another seventeen years.

Even if they had released the information, there would have been another problem. The Russian satellite was broadcasting at 20 and 40 megahertz, frequencies that the network of American radio tracking stations set up to follow U.S. satellites—Minitrack—could not detect, even though every ham radio operator in the world could hear the satellite's distinctive "beep beep." The more sophisticated Minitrack stations, designed for the more sophisticated task of satellite tracking, could only detect signals of 108 megahertz. This was the frequency that the American satellite designers had adopted as the optimum, given available technology. Changing Minitrack to locate the Soviet signals involved far more than twiddling a tuning knob on a radio set.

In the days following the launch of *Sputnik,* Western newspaper stories speculated that the Soviets had chosen the frequency purely for propaganda purposes. Early histories quote some American scientists as saying that perhaps the Soviets chose these frequencies because they had not the skill to develop electronics to operate at higher frequencies.

Yet at the rocket and satellite conference, before the launch of *Sputnik* had colored everyone's view, the Soviets' choice of frequencies was considered in the context of science. One American delegate pointed out that the lower frequencies were better for ionospheric studies, though less good for tracking. A British delegate put forward a proposal for an ionospheric experiment with the Soviet frequencies, which found unanimous backing. During the conference, the Soviets explained a little about their network and specified the type of observations they would like other nations to make.

None of which, of course, means that propaganda did not contribute, but it shouldn't be forgotten that in addition to the political environment, there were also Soviet scientists and engineers with scientific agendas of their own.

Whatever reason or combination of reasons the Soviets had, Minitrack could not acquire and track the satellite. It was to be a week from that night before the first of Vanguard's radio tracking stations was successfully modified to receive the Soviet frequencies. The data, however, were not good, and the American scientists voiced their frustration to one another the following January as they prepared for their own satellite launch.

In the meantime, the phone at the IGY's headquarters rang again. The caller had seen lights in the sky; could they be the satellite? At the IGY they didn't know, but it seemed unlikely. For all of the men at the IGY's headquarters, the intensity of the public's response was a shock. Many of the calls, with their conflicting reports, were a hindrance to the task of determining where the Soviet satellite was and whether the orbit was stable.

Eventually, they realized that the best information was coming from the commercial antenna of the Radio Corporation of America near New York. Engineers there recorded a strong signal at 8:07 P.M. and again at 9:36 P.M. Pickering and his colleagues considered this information, assumed a circular orbit, and then calculated a very rough orbit from the time between signals. They concluded that it was stable. They wrote a press release, then, realizing they'd made a basic mistake in their calculations, recalculated and rewrote. They finally fell into their hotel beds at eight o'clock the next morning. Years later, after numerous glittering successes in space science, Pickering still wonders why he didn't spot the significance of the RCA data sooner. He wonders, too, about the basic error they made, one that was too simple for this group to have made. Yet how could it have been otherwise when they had lost a dream?

Today, when military tracking equipment can locate an object the size of a teacup to within centimeters, when some antennas are set up to follow potential incoming missiles, and track across the sky at ten degrees a second, it is hard to imagine the situation the American scientists faced that night.

Had the satellite been American, elaborate plans for acquisition and tracking would have kicked in. These plans included both optical and radio techniques. Optical because, reasoned many scientists, satellites are heavenly bodies, and who better to track a heavenly body than an astronomer. Radio tracking because this was the obvious next technical development. Radio could work at any time of the day or night, unlike optical tracking, in which observations could be taken only at dawn or dusk in a cloudless sky. For both radio and optical techniques there was to be a reiterated cycle of observation and prediction, gradually refining the orbital calculation.

The scientist knew that the initial acquisition would be difficult because of the anticipated inaccuracies with which the rockets would place the satellites in orbit. The Vanguard team calculated that there would be an inaccuracy in the launch angle of perhaps plus or minus two degrees. Thus, there could be a horizontal position error in a 300-mile-high orbit of about 150 miles at any time. Added to this, the satellite would be traveling at an average of 4.5 miles per second. Further, the anticipated inaccuracy in the satellite's eventual velocity would be equivalent to plus or minus two percent of the minimal velocity needed to stay in orbit. These errors would change a nearly circular orbit into one with some unknown degree of ellipticity.

For the sake of comparison, today's Delta rockets can place a satellite into a low-Earth orbit with a horizontal accuracy of a little under four miles. The angle and velocity of the launch vehicle's ascent to the point where the satellite will be injected into orbit are worked out in preflight computer simulations. Inertial guidance controls monitor the ascent, making whatever angular corrections are necessary to the path to orbit. Such was not the case in the 1950s.

In December 1956, Pickering had told the IGY's planners that the problem they faced was whether they would ever see the satellite again once it had left the launch vehicle.

But of course, the planners worked hard to solve this problem. Minitrack would acquire the satellite, and later Minitrack observations would be complemented by optical observations.

The Vanguard design team at the Naval Research Laboratory took on the radio work under the leadership of John Mengel. Mengel's group used a technique known as interferometry. Several pairs of antennas are needed for this technique, and the distance between each pair depends precisely on the frequency that the array is to detect. Because new positions for the antenna pairs had to be surveyed, it took a week to prepare some of Vanguard's tracking stations to pick up *Sputnik*.

As soon as Mengel heard of the launch, he and his experts on orbital computation set out for the Vanguard control room in Washington D.C. He ordered modifications to the Minitrack stations so that they could receive *Sputnik*'s signal. These stations were located along the eastern seaboard of the United States, in the Caribbean, and down the length of South America. Within hours, additional antennas were on their way to Minitrack stations. The technicians at these sites worked round the clock, improvising in ways they would never have dreamt of the day before. In the Vanguard control room in Washington D.C., others were beginning a seventy-two-hour effort to compute the Russian satellite's orbit from observations that were far less accurate than what they would have had, had their network been operational. The Vanguard team had planned to conduct that month the first dry run of their far-flung network's communication links. Now Minitrack was getting what an Air Force officer called "the wettest dry run in history."

Nearly everyone agreed that radio interferometry would be the best way to acquire the satellite. But radio techniques with satellites were unproven. The transmitters might not survive the launch or might fail. Nor were radio techniques as accurate as optical tracking. Optical tracking was the job of the Smithsonian Astrophysical Observatory (SAO) in Cambridge, Massachusetts. Under the leadership of Fred Whipple, the SAO had plans both for acquisition and tracking. Hundreds of amateur astronomers around the world were to be deployed to find the satellite (the Soviets were involved in similar efforts as part of the IGY). The amateurs' observations would allow the computers at the Smithsonian Astrophysical Observatory to make a crude prediction of the satellite's course, but a prediction that was precise enough for the precision camera, specially designed by James Baker, a consultant to Perkin-Elmer and Joseph Nunn, to be pointed at the area of the sky where the satellite was expected to appear. These precision cameras, roughly the same size as their operators, would photograph the satellite against the background of the stars. The satellite's position would then be fixed by reference to the known stellar positions also recorded in the photograph.

While Berkner was toasting Sputnik at the Soviet embassy, Fred Whipple was on a plane from Washington to Boston. He had been at the conference on rockets and satellites and was on his way home. When Whipple boarded his plane late that afternoon, there was no artificial satellite in space. But satellites can't have been far from his mind. Perhaps he thought fleetingly of the gossip among the American scientists about Soviet intentions. From this thought, it would have been an easy step to recall the previous day's meeting of the United States IGY's satellite committee. They'd tackled the vexed question of delays in production of the precision tracking cameras. Perkin-Elmer was fabricating the optics for these cameras, while Boller and Chivens in Pasadena were building the camera proper. The press had reported that delays were holding up the Vanguard program. These reports were irritating to Whipple. Vanguard had been held up and would have been irrespective of the cameras. But production of the cameras was also delayed. In fact, as far as Whipple could tell, the cameras would not be ready until August 1958, only four months before the IGY was scheduled to end.

This news had not pleased Whipple's colleagues. Dick Porter had summed up, pointing out that by August, there would be only four months of the IGY left to run. If the satellite program was discontinued at the same time as the IGY packed up, the public was going to get a very poor return on the $3.8 million it was spending on precision optical tracking. They'd discussed at length whether they should cancel the cameras but had finally decided to continue because they believed that the space program would continue. That Thursday, the day before the space age began, the IGY participants seriously considered that the satellite program might be canceled.

The big unknown that Whipple must have pondered was the Soviets. Perhaps he remembered Bill Pickering's remarks during the meeting: "I think we very likely face the embarrassing situation that, say, early next spring, we have one or two cameras and the Russians have one or two satellites. We can live with it, but it would be embarrassing; but I think, nevertheless, it is desirable for us to have cameras as quickly as possible."

As far as Whipple was concerned, the problem was that Perkin-Elmer had not put its best people on the job. During the meeting Porter said that he felt like going up to the plant and beating on tables, but that Whipple had discouraged such a move. Whipple's reaction was one of incredulity. He told Porter that he would now encourage this.

Later that month, when Porter did visit Perkin-Elmer, he found that the company had underbid and was now reluctant to pay for overtime when they expected to lose money on the contract. Porter renegotiated the contract so that the company would break even. Together with the launch of *Sputnik,* this greatly speeded up the camera program.

By October 4, 1957, Whipple was battle-scarred. Besides the frustration of the cameras, he faced budgetary problems in the optical program. He was constantly robbing Peter to pay Paul (not that Paul always got paid on time) and he had a lot of explaining to do to both Peter and Paul. Yet he was excited. There were still things to do. It is plausible that Whipple made a mental note to check how the debugging of the computer program for orbital calculations was coming along.

Whipple knew that even if there were still a few wrinkles in the software, his staff were ready to track a satellite. On July 1, the opening day of the IGY, he had told them to consider themselves on general alert. What he did not know was that all of his preparations were at that moment being put to the test. No one enlightened him at Logan Airport, but when he got home his wife was waiting on the doorstep. Within minutes he was on his way to the Observatory.

That evening it looked as though optical acquisition was going to be more important than had been anticipated. Clearly, there could be no radio acquisition and tracking for the time being. RCA's commercial antennas gave enough information to establish that the orbit was stable, but not enough to do any useful science or to predict the orbit with enough accuracy for aiming the precision cameras. Admittedly there were no precision cameras yet, but that was about to change.

Whipple arrived to find Kettridge Hall, which housed the tracking offices, humming with activity. At some point during the evening a fire engine arrived because a woman had reported that the building was on fire. Perhaps she thought that some nefarious activity was underway.

The news of the Russian satellite had reached the observatory at six fifteen. Everyone but J. Allen Hynek and his assistant had gone home for the weekend. Hynek was the assistant director in charge of tracking and had worked with Whipple on the tracking proposal that they had sent to the IGY satellite committee in the fall of 1955. He was discussing plans for the following week when the phone rang. A journalist wanted a comment on the Russian satellite. When the journalist had convinced Hynek that the question was serious, Hynek cleared the line and started recalling those staff who were not already on their way back to work. Those who were members of the Observatory Philharmonic Orchestra were still in the building, rehearsing for a concert. They quickly abandoned their musical instruments for scientific ones.

Hynek was particularly keen to reach Donald Campbell, Whipple's man in charge of the amateur astronomers. Campbell, too, had been at the satellite conference but had remained in Washington because he was leaving the next day for a meeting of the International Astronautical Federation in Madrid. Part of Campbell's job was to ensure that all the amateurs were notified when the time came. The amateurs were called Moonwatchers, after the official name for their venture, Project Moonwatch.

Hynek eventually reached Campbell in Springfield, Virginia, about fifteen miles south of Washington, where he was visiting one of the groups of amateurs. That night, they were conducting a dry run to demonstrate their methods to Campbell. Campbell took Hynek's call, then told the assembled group, "I am officially notifying you that a satellite has been launched." They were thrilled to be part of the first group in the world to hear these words from Campbell. Campbell went on to make a few remarks, the coach rallying the team, but someone stopped him and set up a tape recorder to catch what he said. The next morning the Springfield Moonwatchers were at their telescopes before dawn, but they saw nothing, and would not until October 15.

Whipple had wanted armies of amateurs, but he had had to defend the idea against his colleagues' charges that the amateurs would not be sufficiently disciplined. Ultimately, these amateurs provided invaluable information to teams operating the precision cameras. Although the Moonwatchers had not expected to begin observations until March 1958, when the first American satellite was expected to be in orbit, they were well enough organized that night to begin observing. The first confirmed Moonwatch sightings were reported by teams in Sydney and Woomera on October 8.

During the first night, Whipple, like Bill Pickering and John Mengel, tried to make sense of confusing reports, reports that were not the sort that a professional astronomer was used to. Where were the precise measurements of azimuth (distance along the horizon), elevation (height above the horizon), and time of observation? And, of course, the observatory's program for orbital computations had still to be debugged. IBM, which was under contract to provide hardware and software support, came to their assistance the next day, dispatching experts who helped to debug the program.

Early Saturday morning, Whipple received his best observations so far from the Geophysical Institute, in College, Alaska. By nine o'clock Saturday morning, Whipple was ready for a press conference. He was dressed in a sober suit and accompanied by the props of a globe and telescopes. He gave the appearance of a man who knew what the satellite was doing. Of course, by his own standards, he had no idea.

Over that first weekend, Whipple considered the observations: the object seemed brighter than it should be. He called Richard McCrosky, a friend and colleague at the Harvard Meteor Program, and asked whether the Alaskan observation might be a meteor. McCrosky said no, and speculated that the final stage of the Russian rocket was also in orbit. Whipple contacted the Russian IGY scientists in Moscow, who confirmed that the final rocket stage was indeed in space, trailing the satellite by about six hundred miles. The rocket was painted brightly and had the luminosity of a sixth-magnitude star—bright enough to be visible through binoculars. The official nomenclature for the rocket and *Sputnik I* gave them the names "1957 alpha one" and "1957 alpha two." The rocket, being the brighter object, was "alpha one."

Eventually, Whipple concluded that *Sputnik I* itself was probably painted black. Although he was wrong, it is doubtful that any of the amateurs ever spotted the satellite; their observations, about two thousand of them by the end of 1957, were probably all of the rocket. Certainly, the Baker-Nunn precision cameras never picked up *Sputnik I,* though special meteor cameras that McCrosky lent to Whipple until the Baker-Nunns were ready did acquire the satellite on Thanksgiving day.

The observatory published its first information of scientific quality October 14; the document was called *The preliminary orbit information for satellites alpha one and alpha two.* Later the Observatory issued regular predictions of the time and longitude at which alpha one would cross the fortieth parallel, heading north. More detailed information was available to Moonwatch teams so that they would know when to be at their telescopes to make observations.

One of the Springfield Moonwatchers was a teenager named Roger Harvey. On the evening of October 4, he was driving his father's 1953 Ford back from Maryland, where he had picked up a mirror for a ten-inch telescope that he was building for a friend. He was listening to the radio. When he heard about the Russian satellite, he was exhilarated. Someone had really done it—sent a satellite into space. Now, he thought, we'll see some action.

When President Eisenhower had announced that the United States would launch a satellite, Harvey and his fellow amateur astronomers had wondered what it would mean to them. They'd decided that they would establish an observing station on land owned by the president of their local astronomy club, Bob Dellar. Nowadays Springfield is part of the seemingly endless conurbation of Washington D.C. and northern Virginia. Then it was rural and had a beautifully dark sky for observing.

The group had modeled the layout of their observing station on one that they'd seen in Bethesda, Maryland. One weekend, they'd arranged the observing positions in a single straight line, extending on either side of a fourteen-foot high, T-shaped structure. Harvey thought that the T, which was made out of plumber's pipe, looked like half of a wash-line support for the Jolly Green Giant. The six-foot crossbar was aligned with the meridian, with the north-south line immediately overhead. A light shone precisely where the T crossed the upright. Like most of the Moonwatchers, they'd made their own telescopes, each of which had a 12-in. field of view. The fields of view overlapped one another by fifty per cent, so that it wouldn't matter if one observer fell asleep or missed something.

If one of the team was lucky enough to see the satellite, he would hit a buzzer and call out the number of his observing station at the moment when the satellite crossed the meridian pole. Bob Dellar would have his double-headed recorder switched on. One channel would be recording the national time signal from a shortwave radio, while the other channel would record their buzzers and numbers. The T would determine the meridian; they knew their latitude and longitude; the double-headed tape would have recorded the time of the observation accurately; and they could work out the satellite's elevation to within half a degree by measuring the distance between the central light and the point where the satellite crossed the meridian. Thus they would have elevation for a specific time and place. When they made an observation, Dellar would call the operator, speak the single word, Cambridge, and be put through to the observatory. After that it would be up to the professionals.

There had been dry runs. The Air Force had flown a plane overhead at roughly the right altitude and speed to simulate the satellite's passage. The aircraft had trailed a stiff line with a light on the end. It was important that the Moonwatchers not see the light too soon, so the Air Force had taken the rubber cup from a bathroom plunger, threaded a loop through it to attach to the line to the aircraft, and put a small light and battery in the plunger. The practices had worked well. The plane had flown over with its navigation lights out, which, while strictly illegal, was necessary.

When Harvey got home, he was anxious to hear from Dellar. The arrangement was that the observatory would call team leaders with predicted times that the satellite would cross the equator. Dellar would work out at what time the Moonwatchers needed to be at their telescopes. Of course, it would be a little different now, because it wasn't their satellite and the observatory might not have good predictions. All the same, Harvey was ready when Dellar called the next morning. For the next few weeks, Harvey lived at a high pitch of excitement. He felt himself part of history. Even the police seemed to be on his side. When he was stopped for speeding, he told the officer he was a Moonwatcher, and he was sent on his way without a ticket. The Springfield Moonwatchers felt great camaraderie, and no one pulled rank. On cloudy nights, they would swear at the sky on the principle that if they generated enough heat, they would dissipate the clouds.

On the other side of the continent, in China Lake, California, the skies were much clearer and very, very dark—ideal for observing. Florence Hazeltine, a teenage girl, who was later to become one of the first doctors in the United States to use in vitro fertilization techniques, would bundle up against the cold and ride out on her bike to answer the same calls that drew Harvey to Dellar's house. Like Harvey, she was buoyed by her sense of being part of history.

In Philadelphia, sixteen-year-old Henry Fliegel reported to the roof of the Franklin Institute. He wrote in his observing notes:

> "On October 15, I saw with all the other members of the station a starlike object move across the sky from the vicinity of the pole star across Ursa Major to Western Leo. It attained a magnitude of at least zero when in Ursa Major, but then rapidly faded and finally became too dim to see when still considerably above the horizon, disappearing very near the star Omicron Leonis."

The Philadelphia Moonwatchers had seen Sputnik's rocket. When the news hit the papers, sightseers and reporters turned up and sat at the telescopes, sometimes taking the telescopes out of their sockets as soon as anything appeared.

Elsewhere, the initial confusion was beginning to sort itself out. By the fifteenth, the first precision camera was nearly ready to begin operation. Moonwatchers deluged Cambridge with news of sightings. These were of the rocket, but it didn't matter. It was a body in orbit and the teams were honing their skills. Engineers were ironing out the inevitable wrinkles of the Minitrack system.

The space age was underway.

The Space Age

All of us living beings belong together.

—Erwin Schrödinger

What does the space age offer, and what might it yet be? Perhaps it is no more than an age in which new tools and weapons expand our knowledge and ability to trade and fight wars. A glorified Stone, Bronze, or Iron Age, during which our usual activities will be different only in that they extend beyond Earth's atmosphere. Or is the space age essentially different; was the launch of *Sputnik I* the turning point Tsiolkovsky predicted when he wrote of mankind leaving the earth in pursuit of light and space? Not Russians, Chinese, Frenchmen, or Americans, but mankind, building cities together in space, as he advocated in his science fiction book *Beyond the Earth*.

Space clearly has defense and commercial implications. On the other hand, the United States, Russia, Canada, Europe, and Japan are jointly planning an international space station. The beginning of Tsiolkovsky's vision? Perhaps.

From the beginning, the space age has been home to a well-known threesome: science, human exploration (of which the international space station is the most recent example), and the application of science to military and commercial technologies for Earth. One might expect that the first two, science and exploration, would be the aspects of the space age that would lead toward Tsiolkovsky's vision of a unified humanity. But maybe space science and exploration are not so different in the ways that they can influence our outlook than are science and engineering in other arenas of endeavor—the international effort to map the human genome, for example, or all of the exploration that humanity has undertaken to date. Perhaps in the end it will be the third, at first glance the least different and least glamorous aspect of the space age, that will contribute most to an alternative outlook on the world.

Both space science and exploration have caught our attention with the vastness of their aspiration. *Pioneers 10* and *11,* the first spacecraft to be sent to study the outer planets, have done their job. *Pioneer 10* is now racing down the sun's magnetotail, heading for the interstellar medium and away from the galactic center. *Pioneer 11* is heading for the interstellar medium with the galactic center lying beyond. (William Pickering and the Jet Propulsion Laboratory, incidentally, contributed significantly to these early successes of NASA.) The whole was grandly conceived and has since been surpassed by spacecraft with even grander ambitions.

The *Pioneers* each carry plaques with drawings of a man and a woman, showing their size with respect to the spacecraft. There is a drawing of a hydrogen atom (intended to show our familiarity with the most abundant gas in the universe, but also—unintentionally—a symbol of one of our more devastating weapons). Two other drawings give the spacecraft's path through the solar system from Earth and show our sun's position relative to fourteen pulsars—messages launched from a remote island in space to unknown recipients who may never receive them and, if they do, may not understand them. The urge is familiar, as is the spirit of that blithe inclusion of a return address and the need to believe that the addressees, if they are in a position to respond, are essentially benevolent.

What might that expectation of benevolence be based on? Humility in the face of eternity? "Eternity ... like a great ring of pure and endless light"; the awe expressed in Henry Vaughan's lines written three centuries ago appeared on the faces of the mission controllers in Houston as they gazed at the pictures that the Apollo spacecraft had relayed to Earth of Earth.

Here was form for a poetic metaphor. Yet the view of Earth against the blackness was so spectacular that it has itself become a metaphor.

Science and exploration cannot sustain poetic awe in this or any other age, for all their glamor and beauty.

So what does the application of space technology to solving earth-bound concerns have to offer? When men looked to Earth (women were, for the most part, still waiting in the wings in 1957) and asked what value the space age might have, they thought about tasks they had thought about for millennia: among others, navigation, weather forecasting, and communication—enterprises that in the tradition of previous ages improve the quality of life and facilitate warfare. The hilltop fire flashes news of a battle or of the birth of a child. The general and the farmer have always wanted the weather forecast. Both the master of a merchantman and the captain of a nuclear submarine benefit from better navigational aids.

Of the men and few women who did these things, some were more brilliant than others. Some worked with passionate belief or fascination, others to pay the mortgage. Some had an eye to the main chance, aware that there was money to be made, reputations to be built. Most, doubtless, reconciled more purposes than one. Nor is it possible to say who held what motives in what proportion. At the best of times, the motives of others are difficult to discern and classify. Across time, in a different world, the task is almost impossible. Certainly those in America believed in the importance of their work to the welfare of the United States of America.

The world of 1957 gave good cause for such an outlook. When James Reston interviewed Nikita Khrushchev for the *New York Times* after the launch of *Sputnik,* Khrushchev's speech was littered in all seriousness with descriptions of Westerners as reactionary bourgeois and imperialist warmongers. The background noise included Korea, the Suez crisis, the Hungarian revolution, hydrogen bombs, and advertisements for nuclear shelters in suburban backyards. The searing images then were of the Holocaust and of atrocities in China.

The memory to be lived with and the crucible that formed the participants and in which relationships were forged, was the Second World War. Nearly every nation on Earth was involved. Pearl Harbor had been an unimaginable shock to the American psyche, and the horrors of Hiroshima and Nagasaki were known but not fully realized. Some Americans saw those atomic bombings mainly as a reprieve from witnessing further horrors in the Pacific.

Against this background, when Vietnam, with its legacy of doubt was a thing of the future, America developed a determination to keep the peace through military and economic strength. In defense laboratories, university departments, and industry, scientists and engineers developed satellites that would improve navigation, weather forecasting, and communication. Each now has its place in everyday civilian life as well as in defense.

The military application of providing more accurate positioning for nuclear submarines was the impetus behind the development of navigation satellites. Today, there are more civilian than military users of space-based navigation. This trend began with Transit, the long-lived first generation of navigation satellites. A similar duality exists in the history of communication and weather satellites. Ostensibly, commercial and military applications were developed separately, but the scientists and engineers working on civilian satellites often worked on military projects as well. There was an inevitable cross-fertilization of ideas.

These satellites, pointing to the earth, were truly earthbound in their conception and inception. They were rooted deeply and consciously in defense and commerce and the competition of nations—no transcending idea of mankind in pursuit of light and space. Yet unexpectedly, and in practical ways, these technologies are building from the messy foundations of confused human motives a picture of the earth and its inhabitants that is harder to dismiss in daily life than are the inspirational views revealed by Apollo. Wonderful though that inspiration is, the mundane application satellites are beginning—only beginning—to encourage a practical appreciation of one Earth.

The hurricane that devastates the eastern seaboard of the United States begins as an innocuous atmospheric disturbance over Africa. Navigation satellites can be used worldwide. Satellites make communication possible with places landlocked among political enemies (as in some African countries) or from war and disaster zones that we might otherwise be able to ignore. Faced by the reality of global physical phenomena as revealed by the unique bird's-eye view of satellites, international organizations have sprung up to manage satellites. At the height of the Cold War, ideological enemies cooperated with varying degrees of amity within groups like the International Telecommunication Satellite Organization and the World Meteorological Organization.

Thus these inward-looking satellites offer more than we have yet realized. They are for the first time, and in a very practical sense, a technology that can be fully realized only by considering the earth as an interconnected whole. On October 4, 1957, the first step was taken. Later, as the technology of navigation, weather, and communication satellites evolved, it became clear that the greatest gains or advances in knowledge would come from a holistic view of the world. Of course, the knowledge gained can still serve confrontational purposes. Yet, irrespective of our motives, we see that the nature of the technology itself urges cooperation rather than confrontation. Cooperation might become a habit that sustains the promise inherent in Apollo's luminous images of a blue-green earth.

1

$\star$

Navigation

Polaris and Transit

I don't want any damn fool in this laboratory to save money, I only want him to save time. The final result is the only thing that counts, and the criterion is, does it work?

—Merle Tuve, speaking of the development of the proximity fuse as quoted by J.C. Boyce in *New Weapons for Air Warfare*.

Though *Sputnik I* was a shock to America, it also furnished the United States with the means to develop a technique that allowed the Polaris submarines to aim their nuclear missiles at Soviet cities with greater accuracy.

This is not what two junior physicists, Bill Guier and George Weiffenbach, thought they would be doing when they went to their jobs at the Applied Physics Laboratory on the Monday after *Sputnik* was launched. They tuned into the satellite's signal out of an interest more akin to Roger Harvey's and Florence Hazeltine's than to Fred Whipple's or Bill Pickering's. They had no research aims in mind.

Within days their attitudes changed as they tried to characterize the satellite's orbit as precisely as possible.

Guier and Weiffenbach were among many scientists worldwide who were attempting to determine *Sputnik*'s orbit. All but Guier and Weiffenbach calculated the orbit on the traditional basis of finding angles to the orbiting body. The Moonwatchers recorded what they saw through their telescopes; Minitrack allowed angular measurements to be calculated from radio interferometry; and the radar dish at Jodrell bank, which would normally have been trained on radio sources in the galaxy, swiveled to keep the satellite in its focus, thus giving direct angular measurements.

By contrast, Guier and Weiffenbach recorded the way that the radio frequency of the satellite's transmitter appeared to change as it passed within range of the lab (the Doppler shift). Pickering's committee at the IGY had considered and dismissed the possibility of calculating orbits from Doppler data, concluding that the results would not be accurate enough. Given the techniques of data analysis the IGY committee envisaged, this conclusion was correct. However, Guier and Weiffenbach were to develop an alternative interpretation of the Doppler data, one that was subtle and that relied heavily on new computational and statistical methods.

Those who break with tradition or with accepted modes of practice often provoke hostility, and Guier and Weiffenbach were no exception. Their methods were ridiculed twice: first when they calculated an orbit for *Sputnik,* and then when their method served as the basis for unraveling some of the secrets of the earth's gravitational field.

Fortunately for them, not everyone was skeptical. Among those impressed by their work was their boss, Frank "Mack" McClure, who saw how their methods could provide the basis of a satellite navigational system that would serve the Polaris submarines. That system—Transit—also became the first civilian global satellite navigational system.

Before Transit, the only system of navigation effective worldwide was the one that had been available to Odysseus when he escaped from Calypso and "used his seamanship to keep his boat straight with the steering oar." Odysseus navigated by the stars, keeping the Great Bear to his left and his "sleepless eyes" on the Pleiades.

During the next three thousand years, with the introduction of increasingly sophisticated mathematics, nautical almanacs, accurate clocks, compasses, and instruments to measure angles to the sun, moon, stars, and planets, celestial navigation grew in sophistication and precision. By 1975, when William Craft (who later became a commander and the director of seamanship and navigation at the U.S. Naval Academy in Annapolis) took sightings from the deck of his cruiser, he could determine with a high probability that his ship was within a given circle of radius two nautical miles. By then, says Craft, the practitioners of celestial navigation had become high priests of a secret sect, performing their rites at dawn, noon, and dusk. Though Craft's methods were more elaborate than those of Homer's description, the idea was the same as that which guided Odysseus home—to navigate by the stars.

If Homer had written his epic five hundred years later than he did, Odysseus would have known to calculate his north-south direction from readings of the sun's altitude at noon. Two hundred years after that, by the second century B.C., Homer could, with some loss of poetry, have described Odysseus's journey in terms of latitude and longitude, the system then suggested by Hipparchus for defining position on the earth's surface. However, like mariners of later centuries, Odysseus would have been very unsure of his position when clouds obscured the skies.

Dead reckoning would have helped, particularly in the Mediterranean, where currents are light. Knowing the point of departure, the ship's initial speed (estimated) and heading, mariners learned from the sixteenth century onwards to calculate a new position relative to their starting point. They estimated (by eye) the effects of wind and current. By repeating the process for every tack, that is, by adding velocities (the speed and direction of every tack), and plotting them on paper, they navigated to their destination, making estimated course corrections as they went. Much easier in the description than in practice.

As the journey progresses, the errors add up. If you are wrong in your first estimate of speed and position, clearly you will be even more wrong after the next tack. You might, of course, be lucky and have the errors cancel one another. But how would you know? In clear weather on calm seas, one of the high priests of celestial navigation could check the position estimated by dead reckoning against a position fixed with reference to the stars. By the 1950s, sextants fitted with infrared filters permitted readings through light cloud cover of the sun's position at noon. Nevertheless, sightings were not always possible. Navigators have stories of many days passing during which they could not get a celestial fix. In that time dead reckoning could lead a ship far off course, making steering an optimum course across the oceans difficult and adding danger and cost to the journey.

In the second half of the twentieth century, radar improved navigation in coastal waters. Since the early 1980s, a system known as Omega has provided worldwide accuracies akin to those of celestial navigation, if one knows one's position roughly to start with. But in 1957, when out of range of coastal radar, navigation still came down to dead reckoning and celestial navigation. Not a good system if for some reason you need to know your position as accurately as possible at any time of the day or night, fair weather or foul.

Yet an accurate position fix was exactly what the Polaris submarines would need should the order come to fire, and it was exactly what the submarines did not have. Thus the Polaris development, as a select few members of senior staff at the Applied Physics Laboratory knew, was vulnerable to critics both within the Navy and in other branches of the armed services.

Polaris carried intermediate-range ballistic missiles, capable of travelling 1,200 to 1,500 miles, and was part of the U.S.'s strategic triad of land-based and submarine missiles and long-range bombers. The submarines were deployed in the Arctic and were intended to deter the Soviet Union from launching its own nuclear missiles, the idea being that the United States would always have the capacity for massive retaliation, thus nullifying the traditional military benefits of a surprise attack. Amidst widespread and intended publicity, consistent with the Eisenhower administration's "New Look" strategy, the first Polaris missile was fired from the USS *George Washington* in July 1960.[1]

To achieve its aim, the Navy (specifically, the Special Projects Office of the Navy, later Strategic Systems Projects Office) needed to be certain that its missiles would land within a particular radial distance of the target.

The missiles followed ballistic trajectories. Like bullets from a gun, they were carried to their destination by momentum and gravity. Slight course corrections were possible during the ascent while the solid rockets fired, but these corrections were like those needed to steer down a particular road; they did not permit you to change roads. Accurate positioning and targeting were crucial for getting on the right road.

Although the submarines needed an accurate reckoning of their position, their strength was (and is) in their stealth, constantly moving underwater, prepared to fire at any time and to move away. As long as they remain underwater they are virtually undetectable, and their nuclear fuel permits them to stay submerged without refueling for many months. A force of such submarines is practically invulnerable to a first strike. In this scheme of things, sitting on the surface taking a position fix from the Sun, Moon and stars was not an option.

The Special Project Office's initial solution was that Polaris would carry what was then the comparatively new technology of inertial navigation systems. These, effectively, are automated dead reckoners. They were originally mechanical and are now electronic. They compute from measured accelerations the actual course of a ship, submarine, or missile. They do not refer to anything outside themselves but measure the accelerations experienced in each of three dimensions on the parts they are constructed

1. In October 1953, President Eisenhower's National Security Council endorsed a policy dubbed the "New Look." This policy's aim was that the United States should seek obvious strategic superiority and use rhetoric indicating a willingness to use it. The thinking was that such a policy would deter Soviet aggression and return the diplomatic initiative (post Korea) to the U.S. and permit lower budgets. Eisenhower and his advisors had determined that lower military spending was necessary because the levels at the time endangered national security as much as did inadequate arms. From . . . *the Heavens and the Earth: a Political History of the Space Age,* by Walter McDougall. Basic Books (1985).

from. As such, if the errors can be kept to a minimum, they are ideal for submarines, which may be out of port for some time and rarely surface.

Like dead reckoning, though, inertial navigators accumulate errors. It is difficult to say what those early errors were in Polaris because the targeting and positioning errors of the system are still classified, even though the American Polaris submarines came out of service in the mid 1970s (the last of the British Polaris submarines was decommissioned in 1996). The errors were probably less than those of an inertial navigator on a surface ship, where waves would add to the measured accelerations.

Some of an inertial navigation system's errors were a consequence of the difficulty of manufacturing the mechanical versions with sufficient precision. However, within the Polaris program, anything that money and expertise could have done to improve manufacturing techniques would have been done or attempted, because Polaris (and by extension Transit) carried the naval designation "Brickbat-01," signifying that the project had a high procurement priority. If a computer was waiting to be shipped to someone else and a Brickbat-01 project needed it, the Brickbat-01 project got the computer.

Nevertheless, the accuracy of the inertial navigation system was not good enough. In one day, they accumulated errors that were far greater than those specified for Polaris's positioning accuracy. Since the submarines were on station for months at a time, this was a problem. What was needed was an external reference system that minimized the submarines' exposure; something that would correct inertial guidance errors in the same way that a celestial fix corrects errors in dead reckoning. Such a system was the raison d'être for the Transit satellites and was made possible by the techniques developed by Guier and Weiffenbach.

Transit was important to the Department of Defense for two reasons: first, for navigation; second, because the satellite orbits revealed details of the earth's gravitational field and thus the shape and structure of the earth and the relative positions between geographic locations. It was this second aspect of the Transit program that led at times to a high security classification. By learning about the gravitational field, military planners knew both the position of Moscow, say, with greater accuracy as well as which course to select to the target and what course corrections were needed, and they didn't want anyone else to know how much they knew.

Given that the Polaris submarines were to be deployed in the Arctic, the best orbit for a navigation satellite was one passing over the poles. In this orbit, a satellite would be visible to the submarines every time the spacecraft passed over the north pole, once every ninety-six minutes for a four-hundred-mile-high orbit. From the beginning of 1964, there was always at least one operational Transit satellite aloft. Transit's appeal was that it worked in any weather and the submarine need only approach the surface once at night and deploy its antenna for ten minutes at most. An important attribute of these antennas was their low radar profile, which was important because radars capable of detecting a periscope's wake were then being developed. The system was passive, and thus the submarine did not need to broadcast a betraying signal in order to get a fix.

Within a few years, when more satellites had been launched, it was possible to get a position fix anywhere in the world at least once every three hours, sometimes as often as once every ninety minutes. For the first time ever, navigators could fix their position with greater accuracy than was possible with celestial navigation and could do so more frequently and in any weather.

The original Polaris system specification called for satellites that would locate position to within a tenth of a nautical mile, or about six hundred feet. APL's scientists say this accuracy was available from December 1963, after the first operational satellite was launched. That accuracy improved by the mid 1980s to twenty-five feet. Accuracies of sixteen hundred feet were more typical for cheaper Transit receivers, while surveyors located position to within a few feet with observations of more than one pass.

As a navigational aid at sea, Transit was a marked departure from the past. First, it relied on frequency rather than angle measurements, and second, it made a psychological break with the long past of celestial navigation (radar navigation in coastal waters was also contributing to the change). Instead of wielding sextant, charts, and nautical almanacs, the navigator need only look at the value of latitude and longitude calculated by Transit's shipboard computer from the signals transmitted by the satellite.

Once the Department of Defense made Transit available to civilians in 1967, oceanographers and offshore oil prospectors became the first to use the system, followed by merchant fleets and fishing vessels, and finally pleasure boats.

These developments—both technical and marketing—took some time. The U.S. Navy's surface fleet did not make widespread use of the system until the late 1970s. Aircraft carriers were among the first to be equipped; on cruisers such as that on which Commander Craft served, Transit was installed later. Even when Transit was widely available, naval navigators continued to check its output with position fixes calculated by traditional methods.

The first experimental Transit went into orbit on April 13, 1960. Tom Stansill, an early Transit participant who later joined Magnavox, which manufactured receiving equipment, recalls that the very first receivers on Polaris submarines occupied four racks and cost (in today's dollars) somewhere between $250,000 and $500,000. Typically a rack of electronics was six feet high, two feet wide, one and a half feet deep, and held more equipment than one man could lift. By the late 1960s, when sales opened to civilians, a receiver occupied about half a rack and cost between $50,000 and $70,000. There was a breakthrough in receiver technology in the mid 1970s, and the equipment came down in size and in price to $25,000 and has decreased steadily since then.

Now, of course, the Global Positioning System (GPS) of navigation satellites, which is available every minute of the day, has taken over from Transit. The last operational Transit satellite was scheduled to be switched off at the end of 1996. GPS is so accurate that it can detect the sag on an aircraft's wing. Almost daily, it seems, another unconventional use is found for the system, one of the most recent being as an aid to golfers negotiating the perils of golf courses. Ships can carry electronic charts that GPS updates continuously, and the automation is opening the way to the controversial practice of ship bridges operated by one person. As with Transit, civilian users outnumber the military for whom it was intended, and GPS products are becoming major business opportunities.

Unlike Transit, GPS was designed to include aircraft navigation. To get an accurate position including altitude with Transit, two satellite passes were needed, with interpolation of these fixes by an inertial navigation unit. Transit was initially used in this manner by radar picket planes that remained in the air for many hours. However, a high-speed jet needs a much faster acquisition of navigational information. The great advantage of GPS for aircraft is that its radar imaging technique allows aircraft to get a three-dimensional fix without an inertial navigation unit.

But it was Transit that was the first to provide a worldwide, space-based system of navigation in any weather. And it was the Transit team that first encountered the unknowns of designing a navigational system for space.

The Applied Physics Laboratory, which masterminded the project, was well suited to the task. The lab started out in 1942 as an independent contractor for the Navy. It became part of the Johns Hopkins University after World War II, though the university was not initially keen to embrace a laboratory focusing on defense work.

Merle Tuve, one of the scientists who expressed most doubt about whether satellites should be a part of the IGY, was the laboratory's first head. He had been an advisor to Robert Goddard and was an expert on the ionosphere. He was also among those eating the 21-layered chocolate cake at Van Allen's home in 1950.

Tuve's watchwords as head of the lab during wartime were those at the front of this chapter. To some extent, that ethos still pervaded APL in the mid to late 1950s. One of APL's most important wartime developments for the Navy—a proximity fuze—proved valuable for Transit's development. The fuze was a significant advance for surface-to-air artillery, because it detonated a shell when it was closest to the target rather than on impact, thus increasing the artillery's effectiveness. The development was prompted by the difficulty the Allies had in hitting fast moving aircraft and the V1s and V2s.

More than a decade later when Guier and Weiffenbach decided to determine Sputnik's orbit, they turned first to concepts underlying proximity fuses. And when the Transit development began, the skill acquired by engineers building vacuum circuits for artillery proved, Weiffenbach says, invaluable in building components able to withstand launches.

In 1947, APL formed a research center to work on problems in basic research uncovered during the war years. Many of these were centered on aspects of the upper atmosphere that might affect missiles, and the lab was given some of the captured V2s for its work. In tandem with Princeton University, APL became one of the member institutions of the Upper Atmosphere Research Panel, and James Van Allen was the lab's representative until he left to concentrate on basic rather than military research.

After the war, two men joined APL who were to play an important role in the conception and development of Transit. They were Frank McClure and Richard Kershner. Like Ralph Gibson, who would become head of APL, they came from the Allegany Ballistics Laboratory. McClure, who was to have the idea for Transit, became the research center's first head in 1947. Kershner would become the deeply respected team leader who brought the Transit system to fruition.

In early October 1957, when Guier and Weiffenbach tuned in to Sputnik, McClure was spending half of his time at the Navy's Special Projects Office. Kershner, too, worked with Special Projects and was later made responsible for APL's consultative services to Special Projects. Both McClure and Kershner had a close relationship with Captain (later Rear Admiral) Levering Smith, then the deputy technical director of Special Projects. McClure was thus ideally placed to know of Polaris's need for improved position fixing.

By 1957, Gibson had become head of the lab. Having been sounded out as a potential director of Defense Research and Engineering, he was well placed politically to further plans for a navigation satellite. Gibson had not wanted the job, which was the number three civilian position at the Pentagon, but he clearly had influence and contacts which doubtless were exercised in favor of Transit.

Transit was to encompass a truly formidable array of physics (most branches, including, eventually, quantum mechanics), mathematics, computing and technology. Often the whole Transit system was being designed in the expectation, maybe even blind faith, that when needed, the required new technologies (transistors, battery technology, solar cells, etc) would be in place. It was an approach with which Kershner and Levering Smith were comfortable, and which, in the end, proved successful. For more than thirty years Transit satellites have been in orbit. Now, as the twentieth century closes, GPS has taken over.

But—

It is still only 1957.

It is the Monday after the Friday (October 4) when Sergei Korolev watched *Sputnik* ascend. Korolev is preparing *Sputnik II*. Fred Whipple's orbital determination programs are being debugged. John Mengel is supervising frantic modifications to the Minitrack stations. The police will soon catch Roger Harvey speeding. Bill Pickering has returned to the Jet Propulsion Laboratory, where he is plotting deeply with Wernher von Braun. There is no such thing as Transit. Not even a whisper, nor will there be for another five and a half months. First, there are some important steps to be taken.

Heady Days

When interesting things like this come up, many of the most important results come out of people's curiosity, just following up something, and someone over in a corner who is not supposed to be doing anything about it usually comes out with the best answers.

—Homer Newell, science programs coordinator for Vanguard,
discussing *Sputnik* at a meeting of the IGY's
Technical Panel on Earth Satellites, October 22, 1957

When Wall Street opened on Monday, investors scrambled to buy stock in companies connected with missile programs, abandoning other issues and pushing prices to their lowest level in two years.

Two hundred miles to the south in Washington D.C., there was another scramble as the Naval Research Laboratory briefed President Eisenhower and congressional members on the status of Vanguard. James Lay, executive secretary of the National Security council, called the head of the National Science Foundation, Alan Waterman. The status of the U.S.'s satellite plans, Lay said, would be second on the agenda at Thursday's security council meeting, immediately after a CIA presentation about Soviet defenses. Later that day, Lay called Waterman again; discussion of the U.S. satellite had, he said, now moved to the top of the agenda.

As for the general public, the Saturday morning papers greeted anyone who still did not know of the launch with three-deck banner headlines. *Sputnik* squeezed reporting of the drama surrounding desegregation in Little Rock and Jimmy Hoffa's election as president of the Teamsters Union into a corner of the front page of the *New York Times*. By Sunday, nothing short of a declaration of war could have rivaled the satellite's editorial supremacy. And a *Times* editorial stated soberly, "It is war itself rather than any designated enemy against which we must now defend ourselves."

At his home near the Applied Physics Laboratory, Bill Guier tuned in to every radio and TV broadcast. He was thirty-one, a theoretical physicist, and a news junky even in less exciting times. That weekend, Guier was as fascinated as the rest of the country. He had no idea that the satellite would change his future.

Nor did George Weiffenbach, then a thirty-six-year-old experimental physicist. He too learned during the weekend of the launch of the satellite. Weiffenbach recalls that he was not particularly excited. His wife says otherwise and remembers eager calls to friends and colleagues.

On Monday afternoon, October 7, Guier and Weiffenbach unknowingly took their first steps in the development of a new method of orbital determination and prediction.

Orbital mechanics—the behavior of satellites in orbit—lies at the heart of space exploration. Without a detailed and accurate knowledge of the subject, man-made satellites would be useless, because no one would know where they were or were likely to be. And no one would know the time and position at which a satellite instrument made an interesting observation, such as of an atmospheric disturbance with the makings of a hurricane. Yet it is impractical to follow all satellites throughout their orbits. Hence observations are made of part of the orbit and that information feeds into the mathematical relationships that describe behavior in orbit. Once the orbit is determined, the satellite's future behavior—its position at given times—can be predicted. Orbital predictions allow mission planners to determine, say, when the space shuttle should lift off if it is to dock with the *Mir* space station. They allowed the operators of the Transit navigation satellites to calculate exactly where the spacecraft would be at every two-minute interval in their orbits, thus providing navigators with a reliable celestial fix.

All of the groups observing *Sputnik*—amateurs and professionals alike—were tackling the problem of its motion within the framework of classical physics laid down over the previous three centuries by the likes of Johannes Kepler (working with Tycho Brahe's data), Galileo Galilei, and Isaac Newton.

Brahe (1546–1601) was an astronomer who in the days before telescopes designed and built instruments that enabled him to measure the angles to heavenly bodies. His story is well known in the annals of science because he recognized the importance of basing theory on accurate observations rather than philosophical speculation. His observations of the orbit of Mars were invaluable to Kepler (1571–1630) when Kepler was seeking a fundamental description of planetary motion.

Kepler assumed, though this was far from generally accepted at the time, that Copernicus had been correct in asserting that the planets moved around the sun. With this assumption, Kepler explored numerous mathematical models, seeking one that would lead to a description of planetary motion consistent with Brahe's data.

Kepler's belief in the integrity of Brahe's data kept him at his calculations through years of poverty and illness, calculating and recalculating until finally, theory was consistent with observation. Kepler found that to fit Brahe's observations, the planets must move in ellipses—a geometric figure that has two foci; in the case of Earth-orbiting satellites, one focus of the satellite's orbit is the center of mass of the earth. Kepler next described in a mathematical relationship how a body in orbit sweeps out equal areas of an ellipse in equal times. In a third law, he described how planetary orbits relate to one another mathematically.

In reaching these conclusions Kepler was forced to burn his intellectual bridges, cutting himself off from both views of the universe that were accepted in his day. First, by accepting the Copernican assumption, he was no longer in sympathy with the Catholic Church; second, he had given up the Greek view of the planets moving in circles. This second view was particularly hard for Kepler to abandon. Indeed, it was a view that Galileo was never able to abandon.

At one stage, Kepler thought that the orbits might be circular, but when he made calculations based on this assumption there was a small discrepancy with Brahe's data that kept him at his calculations until he finally concluded that the planetary orbits are elliptical. With less confidence in Brahe, Kepler could easily have stopped his laborious calculations before reaching his fundamental insight.

Kepler was ecstatic. Soon after he formulated the last of the three laws that bear his name, he wrote, "Nothing holds me; I will indulge my sacred fury.... If you forgive me, I rejoice; if you are angry, I can bear it; the die is cast, the book is written, to be read either now or by posterity, I care not which...."

Though Galileo (1564–1642) never accepted elliptical orbits, his giant strides in mechanics provided, along with Kepler's work, a solid scientific basis for Isaac Newton (1642–1727). Among other things, Newton discerned and formulated the all-important law that force is equal to mass times acceleration. All of the different methods of observing the satellite's position, Guier's and Weiffenbach's included, sought different ways to gather information that could make use of this law—the so-called second law of motion—within the concept of orbits established by Kepler.

Newton's second law (F = MA) and Kepler's second law allow mathematicians to deduce six numbers that give an approximate description of an orbit. In tribute, these numbers—or elements—are named after Kepler. One gives the orbit's shape, that is, how elliptical it is, whether it is extremely elongated or close to circular. This element, termed eccentricity, is defined as the distance between the two foci divided by the length of the major axis. The closer the value of eccentricity is to one, the more elliptical the orbit. An eccentricity of zero describes a circle.

A second element, the semimajor axis, gives the size of the orbit. Two more describe the orbit's orientation in three dimensions with respect to the celestial sphere, and a fifth pinpoints that orientation with respect to the earth (respectively, inclination, ascending node, and argument of perigee). The sixth element is the time at which the satellite is at the ascending or the descending node, that is, the time at which the satellite crosses the plane of the equator heading either north or south. It is more usual to use the ascending node.

Kepler's elements, however, are only averages. Any additional force, such as an inhomogeneity in the earth's gravitational field, that acts on the satellite causes it to deviate from the average path described by Kepler's elements. The deviation might be small and local or might cause the entire orbit to precess; that is, the orbit can rotate about an axis in such a way as to sweep out a conical shape. Thus a description of a satellite in near-Earth orbit requires eight parameters: the six Keplerian elements and two precession terms. While such a description is more accurate than that provided by the Keplerian elements alone, it is still only an approximation. Only by giving the positional coordinates at given times can an accurate orbit be described, and this is what the Transit team eventually did.

On Monday, October 7, people were still struggling simply to observe *Sputnik*. The Minitrack stations were being adapted; only some of the Baker-Nunn cameras were in place; the first confirmed Moonwatch sightings, in Woomera and Sydney, Australia, would not be made until the next day.

Across the ocean in England, at Jodrell Bank and the Royal Aircraft Establishment, radio telescopes were waiting for *Sputnik*. They tracked the satellite by keeping its signal at the focus of their dishes, and the angle of the telescope relative to a fixed north-south line and to the horizon gave respectively the azimuth and elevation of the satellite. These data can be used to determine the Kepler elements. When in April 1958 Guier and Weiffenbach published a brief summary of their methods and results in scientific correspondence to the journal *Nature,* they compared the orbit they calculated for *Sputnik I* with those from Jodrell Bank and the RAE. There was good agreement.

Once their technique was developed, Guier and Weiffenbach's orbital determination could, unlike those from Minitrack and the Smithsonian Astrophysical Observatory, be based on observations from a single satellite pass. Eventually, for truly accurate orbital determinations their technique would be applied to the observations collected by many ground stations over a twelve to eighteen hour period. When reversed, the technique allowed position to be fixed at a receiver on Earth from one satellite pass, the basis of Transit.

As Guier and Weiffenbach made their separate ways to work that Monday, all this was shrouded in the future. They knew each other slightly as fellow members of the research center. They had worked together on some of the same projects during the previous few years. By chance, that Monday they ate lunch at the same table in the canteen. Just as at the White House and the Pentagon, *Sputnik* was the focus of conversation.

Guier thought the satellite was superb. Yet his history was not one that obviously suggested an admirer of Soviet ingenuity. As a graduate student in the late 1940s, he had heard Edward Teller, "father of the hydrogen bomb," call for physicists to fight a new kind of technological war. In a world newly afraid of atomic power and rife with the political tensions that would lead to the Berlin Wall, Teller's call to arms seemed logical to Guier. Guier joined APL's research center in 1951.

Like Guier, Weiffenbach believed in the importance of weapons research and in the protection that technological supremacy could afford. He had been in Europe during the war and had been among those waiting in the summer of 1945 for orders to go to the Pacific. On his demobilization, the government paid for his education, something his parents could not have afforded to do, through the provisions of the GI Bill. It seemed right to Weiffenbach to work to strengthen national security, both because of what he had seen in Europe and to repay the nation for his education.

Guier's choice had a consequence that, though small in the scheme of things, was indicative of the times. He had a German friend behind the

Iron Curtain. Their correspondence helped Guier with the language studies necessary for his doctorate. Guier did not see this man through a political lens, and he was staggered when his thesis director advised him to end the correspondence. The Soviets might exploit the relationship, said the professor, and even if they didn't, the letters could damage Guier's career.

At first Guier ignored the advice, but a summer internship at Los Alamos changed his mind. Assigned to work on computer models of nuclear explosions, he learned of the secrecy surrounding nuclear weapons, of the comparative simplicity of some of the physics, and the great fear in government circles that the Soviets would soon have similar technology. These fears were heightened when the Klaus Fuchs and Rosenberg scandals erupted. Fuchs, a British physicist, and Julius and Ethel Rosenberg, an American couple, gave atomic secrets to the Soviets. Fuchs was imprisoned, the Rosenbergs executed. In this climate, Guier became acutely aware of the Cold War and reluctantly ended his correspondence.

Accepting the political realities of their world, Guier and Weiffenbach settled down to conduct basic research in support of weapons systems. The Soviets were the enemy.

And yet . . .

Here was something new, the first foothold in space. The lunch group asserted, with the scientist's classic understatement, that *Sputnik* was not trivial. APL staff, deeply involved in missile and radar work, knew this well. And despite its provenance, *Sputnik* was intriguing.

The lab was not involved with the IGY. Thus people at APL, certainly at Guier's and Weiffenbach's low level, had no inside track to what little was known about *Sputnik*. Nor was there any involvement with the planning for Minitrack. Their ignorance of the IGY's discussions about the limitations of Doppler and the advantages of interferometry freed Guier and Weiffenbach from the conventional wisdom about satellite tracking and orbital analysis and ultimately stood the pair in good stead.

Lunch wound down.

Guier and Weiffenbach decided to rectify the surprising fact that in a lab bristling with antennas and receivers nobody was listening to *Sputnik*'s signal. That was all: an indulgence of curiosity. They did not intend to make a serious attempt to determine and predict *Sputnik*'s orbit, but only to calculate roughly when the satellite would reappear over Washington.

When he got back to his lab, Weiffenbach attached a wire to the back of a receiver and prepared to tune in. He knew—how, given the widespread publicity, could he not—that *Sputnik* was broadcasting an intermittent signal at about twenty megahertz.

Serendipity, that great friend to science, now lent a hand for the first time. Weiffenbach's receiver, working on the same principle as any ham radio operator's equipment, added the satellite signal of approximately twenty megahertz to a reference frequency of twenty megahertz, resulting in an audible "beat" note—*Sputnik*'s distinctive beep beep. But Weiffenbach's reference frequency was special; it was broadcast by the nearby National Bureau of Standards as a service to scientific laboratories, and it was at a precise twenty megahertz.

Some few hours later, when their rampant curiosity turned slowly to a controlled scientific interest, Guier and Weiffenbach started recording the beat frequency along with the national time signal during each pass of the satellite from horizon to horizon. And that was the raw data from which they developed a new technique for orbital analysis and provided the scientific underpinning for a satellite navigational system. Without a precise reference signal, they could not have extracted the information they did. Had the reference signal been broadcast from a more distant site, its quality would have been degraded on passage through the atmosphere, and again they could not have extracted the information they did.

As they waited for the satellite to come over the horizon, Weiffenbach fiddled with the knobs of his receiver. The airwaves were buzzing. Disembodied voices mimicked the satellite's signal. Others repeated, "this is your Sputnik ... this is your Sputnik." Then—faintly—Weiffenbach picked up an English-language broadcast from Moscow. It gave the times when the satellite would be over the world's capitals. They knew when to listen.

By now a crowd had gathered. During the next few days there would at times be so many people in the lab that Weiffenbach could scarcely squeeze through to his receiver. As they waited, there were a few murmurs, not serious, that perhaps the signal was a fake. Most wondered whether the signal would carry information; measurements, perhaps, of the satellite's temperature—the first remote measurements (*telemetry*—measurement at a distance) from space.

The same question arose at a meeting of the IGY's satellite panel on October 22. Inevitably, the main topic on the agenda was a discussion of when America would be ready to launch its own satellite. But the panel also discussed, without reaching a conclusion, whether it could squeeze money from the budget to analyze *Sputnik's* signal for telemetry. Bill Pickering said that even if the signal did carry temperature or pressure measurements, they would be useless without knowing how the Soviet instruments were calibrated. Homer Newell observed that maybe someone unconnected with the IGY's work would decipher *Sputnik's* message. (The Soviets had said, incidentally, that there would be no telemetry from their first satellite.)

The discussion was academic: *Sputnik* stopped transmitting a few days after that meeting. And long before October 22, Weiffenbach and Guier had concluded that there was no telemetry.

In Weiffenbach's lab, the crowd heard *Sputnik's* beep for the first time late on Monday afternoon. Guier, an amateur musician with good relative pitch, cocked an ear. Surely the pitch of the beat note was changing. Of course, he said, it must be the Doppler shift. The change was subtle, a matter of about an octave over twenty minutes. It was this phenomenon, known to every physics student, that was to provide the key to the Applied Physics Laboratory's work on geodesy and Transit.

The Doppler shift is, at heart, conceptually quite easy to grasp. Its consequences can reveal much about the natural world. Astronomers, for example, find it of great utility when studying the motion of stars. But, as with much else in physics, once one goes beyond the basic description of a phenomenon, things become complicated, and the Doppler effect is not easy to exploit in the real world. For this reason, some mistrusted Guier's and Weiffenbach's work. Others were suspicious of their mathematical analysis. The two scientists, recalled Guier, would be told "reliably" what criticism this scientist or another had made of their work and were all but told they were cheating.

A straw poll soliciting descriptions of the Doppler effect elicited comments ranging from the general statement that it has to do with the way the sound of a car's engine changes, to the statement that frequency increases as a source moves toward you and decreases as it moves away. Given its importance to Transit and the understandable vagueness of many, it seems worth describing the basic concept in some detail.

The Doppler shift, which is a consequence of a moving frequency source, is usually introduced to unsuspecting physics students through a discussion of how and why the pitch of a train's whistle changes as the train races past. It is pointed out that the change in pitch occurs even though the pitch of the whistle would not be changing if you were sitting next to it on the train's roof.

I'd prefer to start at the seaside, and to tackle the train analogy later.

Imagine facing the ocean for twenty seconds and counting the waves as they crash into you. Now imagine walking into the ocean to meet the waves head on. More waves will reach you in those twenty seconds than if you had stood still. Now imagine walking out of the ocean directly away from the waves, and you can see that fewer crests will reach you in twenty seconds than if you had remained stationary. So the number of waves generated by the ocean in a given amount of time (the transmitter frequency) remains constant, but the number of wave crests encountered (received frequency) increases or decreases because of your motion relative to the waves. The difference between the transmitter frequency (ocean waves rolling in per second) and the received frequency (wave crests encountered) can be thought of as the Doppler frequency—the amount by which the received frequency differs from the transmitter frequency because of relative motion between a frequency source and a receiver.

Now imagine that the ocean is unnaturally well behaved and that the wave crests are spaced evenly (the transmitter frequency is constant). You are walking into the ocean at a steady pace. In the first twenty seconds you encounter a certain number of wave crests. If your pace and direction forward stay the same, how many wave crests do you encounter in the next twenty seconds compared with the first twenty seconds?

The same number, of course, because nothing about the ocean waves or about your motion with respect to them has changed between the consecutive twenty-second intervals. So the frequency received, which was Doppler-shifted upwards by your initial motion forward, remains constant. A graph of received frequency against time would be a straight line parallel to the horizontal, or x-axis. When you turn around and retrace your steps at exactly the same pace, the frequency received will be Doppler-shifted downwards and will remain constant at that value. A graph of received frequency against time would also be a straight line parallel to the x-axis, but this time at a value below the transmitter frequency. Momentarily, as you stop and turn, you will be buffeted by waves at the natural frequency of the ocean.

Once you make it to dry land, turn back to the waves. They are now behaving in a very peculiar way. The wave fronts are aligned exactly parallel to one another, wave crests evenly spaced. They roll in at a rate of 100 per second. You walk directly into them, along a path perpendicular to the wave front. You are walking at a pace which in one second permits you to cover a distance equal to six wavelengths. So in one second you encounter 106 wave crests. (Transmitter frequency is 100 waves per second, or cycles per second; the received frequency is 106 cycles per second and the Doppler frequency is 6 cycles per second.)

Now instead of walking a path perpendicular to the wave front, walk into the ocean at a slant. The number of waves crests you encounter in a second will still be greater than if you stood still (will be Doppler shifted upwards), but it will not be as great as if you took the perpendicular path. The received frequency and whether or not it is shifted upwards or downwards and by how much depends on the relative motion between you and the wave front. See the diagram below.

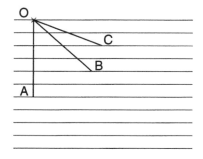

The same logic can be applied to the sound waves of a train's whistle and to radio waves from a satellite.

Let's take the train. If you and the train remain stationary, the number of sound waves emitted by its whistle in a given time is the number that you hear in that time, just as the number of waves rolling ashore in a given time interval was equal to the number of wave crests that hit you when you stood still.

If you are standing on the track and the train is moving directly towards you at a constant speed, the rate at which the range between you and the train changes remains constant just as when you were moving into the water along a path perpendicular to the wave front at a pace of six wavelengths per second. So the frequency received from the train's whistle is shifted upwards because of forward motion and remains constant at the new value as long as that relative velocity (speed and direction) remain constant. Then as the whistle moves through you, for a moment you hear the natural frequency of the whistle (the Doppler shift is zero). Then the train moves away and the frequency is Doppler shifted downwards, and that Doppler shift also remains constant while the relative velocity remains constant. This is all logically equivalent to what happens as you walk in, turn, and walk out of the ocean.

Now what happens if you are standing off at some distance on one side of the track? The rate at which the distance between you and the train changes is no longer constant. That is, the relative velocity between you and the source is no longer constant. So the Doppler shift is no longer constant.

The diagrams below, where marks on the track represent the train's position at, say, fifteen-second intervals, show how the rate of change of distance between train and observer changes with time, and how that change depends on how close the listener is to the track.

Again there is a logical equivalence between walking into the ocean at a constant angle or changing the angle as you walk into the waves. Thus, the amount by which the Doppler frequency is shifted above the transmitter frequency *decreases* gradually as the train moves towards you, until at the point of closest approach the received frequency is affected neither by the forward nor retreating motion of the train. Afterwards, the frequency continues to decrease.

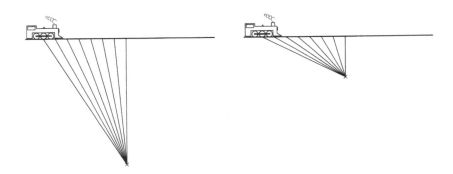

Graphs of received frequency against time would look like those in the next two diagrams.

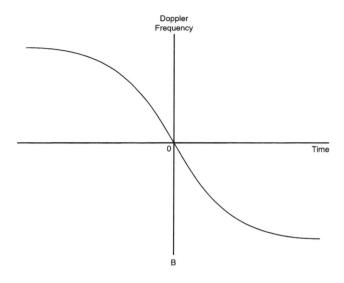

The observer is further away from the track in the example below compared with the one above. If you look at the way distance is changing with
time in the train diagrams, that is, at the way the relative velocity changes
with time, the graphs make sense. This relative velocity, on which the
Doppler shift depends, changes less the further away the moving source
remains from the listener. Hence the change in frequency is less pronounced in the diagram below.

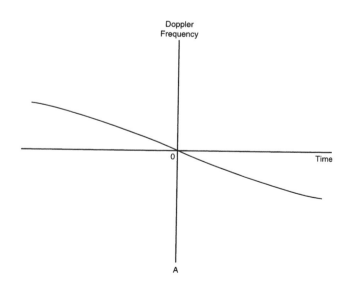

Sputnik was in motion, so its transmission was Doppler-shifted. Unlike the train whistle of our example, however, the frequency arriving at the lab was way outside the range of human ears. Instead, the received frequency was combined with a reference signal to give a "beat" signal in the audible range. Guier's musician's ear detected changes in the beat frequency as a change in pitch.

Many amateur listeners who heard the phenomenon dismissed it as receiver drift. Guier and Weiffenbach did not, first, because they were physicists listening to a moving frequency source and the concept of Doppler shifting was probably hovering not too far from conscious thought; second, they both knew that they had a good reference source and that Weiffenbach's experimental experience would have enabled him to recognize receiver drift or the consequences of a poor reference source for what it was.

Before Guier and Weiffenbach heard the Doppler shift, they were listening to *Sputnik* out of the same curiosity that drove the Moonwatchers and the rest of the country. Once they recognized the Doppler shift, they became more serious.

As *Sputnik's* beep faded over their horizon for the first time, they decided to record the sound of the Doppler shift simultaneously with the national time signal. Although they had no particular idea of what they would do with the recordings, they knew that those beeps, which were exciting imaginations across the world, would give them the scientist's bread and butter: data. And with application and imagination, data can become information.

Another pass was to occur that night. Guier went home to collect a high-quality tape recorder he had recently bought. Back in the lab, Weiffenbach tuned in well before the satellite appeared. They wanted to collect the complete Doppler shift resulting from a pass from horizon to horizon. During the remaining two and a half weeks that *Sputnik* transmitted, they recorded the Doppler shift every time the orbit carried the spacecraft within range of the lab. And thus the Soviets provided the experimental setup that was the first step to enabling the Polaris submarines to fix their position at sea more accurately and opened the way to a global navigational system.

Serendipity, in the form of timing and human relationships, now made its second appearance. Weiffenbach, the experimentalist, and Guier, the theoretician, found they worked well together. Experimental and theoretical physicists, and this is something of an understatement, do not always see eye to eye. During the coming six months, the two men's skills and outlooks would mesh, gearing up their productivity. In later years as they got to know one another better and were part of the team developing Transit, they would have their offices close together so that they could more easily discuss the problems they encountered.

Sputnik's launch came at the right time for both men. Each was ready for a new project. Guier was feeling a new-found confidence, having emerged recently from chairing an APL committee doing technical work on long-range missile guidance systems for the president's Science Advisory Committee. Weiffenbach had things to prove. He had been scooped twice, once on work for his doctorate.

By Tuesday, Guier and Weiffenbach knew that there was no telemetry, but they had the recordings. With these they intended to calculate roughly when the satellite would next appear over their horizon. To estimate the satellite's next appearance, they decided to apply physics in the way that the lab had when designing the proximity fuze which detonates a shell on its closest approach to a target. All they needed to do was to find the distances at the point of closest approach between the satellite and the lab—the point where the Doppler frequency was zero—during several passes. If they found these distances, they would have enough data to determine approximate values for the Keplerian elements and to predict roughly when a satellite in that orbit would next be at its closest to the lab.

In taking this approach, Guier and Weiffenbach were applying physics in a well-known way, but the difficulties they encountered were to lead to an innovative interpretation of the Doppler data.

Given APL's previous work with proximity fuzes, perhaps it was inevitable that Guier and Weiffenbach should have started as they did. An artillery shell emits a radio signal that is reflected from the surface of the target. At the point when the shell has made its closest approach to the target and begins moving away, the Doppler shift is zero and the fuze detonates the shell. The technique was important, too, to the lab's missile work. Missiles would telemeter the Doppler-shifted radio signals that were reflected from the target back to APL's engineers, who would analyze them to ascertain how closely the missile had approached its target.

Guier's and Weiffenbach's first step, therefore, was to turn their recordings into the tables of frequency versus time needed to plot a graph of the satellite's Doppler shift, because the information from that graph, the Doppler curve, was needed in the equation to find the distance when the satellite was closest to the lab (in missile parlance—the miss distance).

Today, and indeed not too long after Weiffenbach and Guier began their work, automated equipment, the inner mysteries of about which the scientist need not be concerned with, would have done the job. That October, Weiffenbach operated the equipment—a wave analyzer—manually. It was standard equipment, and the natural frequency of the electrical circuits it contained could be adjusted by hand. When a frequency on a recording coincided with that of the analyzer's circuitry, the analyzer would record a spike. All of the recordings were fed through by hand and the circuitry adjusted laboriously until fifty values of frequency and time had been extracted for each pass.

Sometime early in the process, Guier and Weiffenbach had been joined by two engineers, Harry Zinc and Henry Elliott. Zinc had also been keen to listen to *Sputnik*. Unknown to Guier and Weiffenbach, he and Elliott had put together equipment with a moderate-gain antenna that was capable of providing a stronger signal than that obtained by the wire attached to Weiffenbach's receiver.

The four took turns with the tedious job of turning the recordings into tables of frequency and time. An unglamorous task, but an essential part of the scientist's life—data reduction.

Work began in earnest on Tuesday, October 8. Very quickly, things began to look ambiguous and complicated. Eventually, from this confusion came the idea that led the scientists to being able to determine an orbit from a single satellite pass—the idea that, when turned on its head, was the basis for the Transit system.

Pursuit of Orbit

Bill Guier sees an equation as paragraphs of lucid prose. There are nuances and implications stemming from the relationship that the equation establishes between different aspects of the physical world. To George Weiffenbach, numbers, their intrinsic values and relationship to one another, are like a film. How they change tells him how the physical events that they represent are unfolding. These ways of viewing the world are quite usual for physicists, and the one is typical of theoretician, the other of the experimentalist.

Each understood something of the other's outlook and could, to some extent, see from the other's viewpoint. The combination proved invaluable.

Some textbooks still describe the miss-distance method that Guier and Weiffenbach initially adopted as the basis for Doppler tracking. It is not the technique that APL developed. Had Guier and Weiffenbach stuck to determining miss distances, they would not have had tracking data that were accurate enough for a good orbital determination, because one of the values needed in the calculation is satellite transmitter frequency. Sputnik's oscillator, though loud and clear, was varying slightly from its nominal value in an unknown way. Those at the IGY who had discounted Doppler as a means of obtaining tracking data had done so precisely because they were concerned about the stability of the oscillators that generated the radio signals. Oscillator stability would be of concern, too, in the method that Guier and Weiffenbach developed, but their approach offered a way to tackle the problem.

On Tuesday morning, October 8, Guier and Weiffenbach were still thinking in terms of miss distance. And they were conceptualizing the problem. As is usual when creating a mathematical model of a physical phenomenon, they simplified the problem. In this way, they would find out what general principles were at play before turning to the inherent complexities of this particular case. They assumed that the satellite was moving in a straight line, just like our train. Since they were looking only for the range at the point of closest approach, the simplification was reasonable.

Guier and Weiffenbach knew all but one of the values needed to calculate miss distance, and that value—the range rate when the Doppler shift was zero—was available experimentally from the slope of the curve as it passed through $T = 0$ in the graphs of frequency against time. It is the steepest part of the curve.

Guier recalls watching colleagues plotting graphs, fitting plastic splines to them and measuring the slope, then looping string around the lab to represent an orbit and marking the miss distances with rulers. It was for fun as much as anything, though such a physical representation of the orbit does give a conceptual idea of what is going on in three dimensions.

To two people who saw what Guier and Weiffenbach saw in equations and numbers, the ruler and string approach was not the way forward. They determined the maximum slope from tables of frequency and time using a mechanical calculator. Soon, they switched to the lab's new Univac 1103A, APL's first fully digital electronic computer. With its combination of vacuum tubes and transistors, the 1103A could perform several thousand additions per second, zipping through the tables of time and frequency far more rapidly than was possible with a mechanical calculator. It was an advance on APL's partly analog machine—a type of computer that performed arithmetic operations by converting numbers to some physical analogue, say length or voltage.

By the late 1950s, university departments were slowly converting to digital machines, but not all scientists and engineers had, as yet, seen the value of computers. So, even though their project was not sanctioned officially, Guier and Weiffenbach had no problem getting time on the computer, something that would not have been the case even a few months later.

When Guier and Weiffenbach first turned to the Univac, some of their colleagues asked, Why do you need a computer to calculate the slope of the curve? Given that they intended only a rough calculation, the question was reasonable. Guier, however, had worked with computers at Los Alamos. Though those machines were mechanical and rattled and banged around once Guier had fed in the numbers and instructed the machine to add or multiply, he had been won over by them. To him, the 1103A was a luxury, and he was happy to find a problem for the new machine.

It was no easy job, however, to program those early computers. A task that would take an hour to code in a higher-level language such as FOR-TRAN might take eight hours to code in assembly language, where each mathematical operation had to be coded for separately. If two numbers were to be multiplied together, the code needed to tell the computer in which memory locations to find each number, in what location to carry out the multiplication, and where to store the result.

Guier wrote instructions for those writing the code. As the orbital determination grew more elaborate, the number of code writers increased, and it not always possible to tell what each programmer had done. Guier started to insist that code be accompanied by notes describing why a particular approach had been taken, and, though it looked like a tedious task, he learned assembly language himself in order to provide some continuity in the software effort (of course, this all took place before the word software had been invented). Thus APL, like others at the time, started to formalize software development.

Today's computers can translate high-level languages, which are rich in symbolic notation, into the low-level assembly language that the machine "understands." Computers then did not have the memory to hold the programs needed for such a translation. To make matters worse, code writers often had to go back to the basic binary words of zeros and ones, which the limited symbology of assembly language represents, in order to find errors.

Despite what today would seem to be daunting limitations, the Univac 1103A was a godsend to Guier's and Weiffenbach's research.

As the two sought values for the miss distance, they soon recognized the ambiguities inherent in the approach. In the case of the train, it is easy to see that the way the frequency varies depends on how close the listener is to the track. But if you have only a tape recording, how do you know whether that recording was made two yards to the east or two yards to the west of the track? In the two-dimensional case, you are stuck with the ambiguity. In the case of the satellite, the earth is rotating beneath its orbit, and the east-west symmetry is broken.

But there are other ambiguities. Consider the train again; a recorder two yards west of a north-south track will record the same frequency shift if it maintains its westward distance but moves a mile due north. Fortunately, satellites inhabit our three-dimensional world. Rather than moving in a straight line, the satellite's path with relationship to the lab might be a shallow or pronounced arc, near to the lab or like a distant rainbow, low or high on the horizon, and the arcs could be at many different angles. For each arc, the relative motion between the lab and the satellite, and thus the Doppler shift, was different.

Guier and Weiffenbach began to recognize the richness of the situation, its complexity. The miss distance soon ceased to look like a particularly interesting value to calculate. They were beginning to suspect that their Doppler curves might contain a lot more information about satellite motion than was immediately apparent.

So they decided to explore, to find out how much information the data contained. To do this they needed to know how the Doppler data changed for different orbits. But only one satellite was aloft. Guier turned to the computer and created a general mathematical description of an orbit that would then generate the theoretical Doppler shift associated with a hypothetical satellite's motion.

Guier pulled together the geometry, trigonometry, and algebra required to describe a generalized ellipse in space and the relationships between the ellipse, the lab, and the center of the earth. He drew up flow charts showing how his mathematical description of this physical phenomenon could be turned into a computer program.

The model assumed the earth to be spherical, which it is not, and orbits to be circular, which they are not, though Sputnik's orbit was nearly so. The idea, again, was to deal first with the simplest possible situation in order to clarify the underlying principles. This construction of a generalized mathematical representation is akin to preparing a canvas for an oil painting. Guier's and Weiffenbach's painting would prove to be ambitious.

They began to generate theoretical Doppler curves. Of course, they were not creating curves but churning out lists of Doppler-shifted frequencies and times for the computer to work through, looking for ways in which the Doppler data varied as the orbital path varied. These numbers were octal rather than decimal, so instead of a counting scheme based on the digits 0 to 9, the digits run only from 0 to 7.

Guier and Weiffenbach pored over these numbers, and as necessary turned to their mechanical calculator to convert octal to decimal. They decided to make their model more realistic and so included mathematics describing the earth's oblateness (a bulging at the equator caused by the centrifugal force resulting from the earth's rotation). Oblateness causes orbits to precess because the Earth's gravitational field is not uniform, as it would be around a homogeneous sphere. Inclusion of a term for oblateness was the first brush stroke on the newly prepared canvas.

In 1957, geophysicists thought they had a good understanding of the earth's shape and structure and thus of its gravitational field. They turned out to be very wrong. Observations of satellites in near-Earth orbit showed all kinds of deviations from Keplerian orbits, each one representing some variation in the gravitational field, which in turn was a result of inhomogeneities in the earth. The field of satellite geodesy was established to tease a new understanding of the earth from these observations of satellite motion. In 1957, the revolution that satellite geodesy precipitated in our understanding of the earth had not yet begun, and the mathematical model that Guier devised to generate orbits was extremely rudimentary compared with what is possible today. In the years following Sputnik's launch, observations of satellite motion would lead to many more brush strokes on Guier's mathematical canvas, each representing a newly understood aspect of the Earth's structure and its associated gravitational field.

Using their basic model in October 1957, Guier and Weiffenbach generated more theoretical orbits. They changed imaginary perigees, then the inclinations, ascending nodes, and eccentricity, and they observed what impact these changes in the Keplerian elements had on the theoretical Doppler curves generated by the computer. By this stage, they had completely lost interest in calculating miss distances.

Given the multiplicity of tasks and the extent to which they exchanged ideas, it is hard now for Guier and Weiffenbach to remember who did what when, but they agree that at some point in those first days, Guier recognized that *the detail of the Doppler curve must change in a unique way, depending on the geometry of the satellite pass.* This was the conceptual breakthrough that others failed to see or, if they saw it, did not use: that each ballistic trajectory of the satellite generates a unique Doppler shift. They asked, Could they find the Keplerian elements that would generate theoretical Doppler curves that matched experimental observations of a real pass? If they could—bingo! The Keplerian elements fed into the theoretical model would be the elements defining *Sputnik's* orbit. They could test their findings by predicting the time that the real satellite would be at a given position if it had the Keplerian elements they had found by this method.

They knew enough from their observations and knowledge of physics to make rough initial guesses about *Sputnik's* Keplerian elements. They fed these to the computer, generating theoretical lists of Doppler shifts. They watched the output, the reams of octal numbers, looking at how closely the theoretical values matched the observations for a particular pass.

Their project was still not sanctioned officially, but other members of the research center knew that something potentially interesting was afoot.

Their boss, Frank McClure, always insightful if not always diplomatic, was watching—at a discreet distance. From time to time, the trademark pipe of Ralph Gibson, APL's director, appeared stealthily around a corner.

At some time in the first few weeks, when things happened in a less orderly fashion than this account suggests, Guier heard that Charles Bitterli, one of APL's few computer programmers, was writing an algorithm for a standard statistical tool, known as least squares fitting, that would be very useful in Guier's and Weiffenbach's research. Guier went in search of Bitterli.

Bitterli, who needed a real-life test problem, was happy to oblige by making his work available. An algorithm is a step-by-step set of instructions for carrying out a computational process. When translated into a computer program, it becomes a standard tool to be pulled out when needed. Today, thousands exist in software libraries. Back then the task of building such libraries had scarcely begun.

Guier's and Weiffenbach's problem was that vast numbers of theoretical Doppler curves could be generated by varying the Keplerian elements. If it were not for the fact that the physics ruled out some combinations, the number of options would have been impossibly large. Even with the limits that physics imposed, there were many possibilities.

Each theoretical curve had to be checked point by point against observations of *Sputnik*'s Doppler curves. One theoretical curve might at first look like the best match, but then a comparison point by point could well show that it was not. Further, once they found the best possible match, they needed some quantitative way of assessing how good that fit was, and thus what the errors in the Keplerian elements probably were. So an algorithm for least-squares curve fitting, a statistical technique for finding the curve that best represents a set of experimental data, was very desirable to Guier and Weiffenbach. Bitterli's program, which was for linear equations only (those whose graph is a straight line), had to be generalized for nonlinear equations relating many parameters. It was now that it became apparent that new methods would have to be learned and that specialists in numerical analysis would be needed.

Bitterli's algorithm worked well, and the comparison of theoretical and observed curves speeded up. Even so, and despite the fact that their intuition was working full time as they made informed guesses about the input values of the Keplerian elements, they couldn't generate a match that was close enough to be of any help in defining *Sputnik*'s orbit. Yet, theoretically, the method should work.

What next? Check everything; examine the setup for recording the Doppler shifts—connection by connection. Review the raw data, the calculations. Weiffenbach combed through everything he could think of. He concluded that the data were good. Guier reviewed the theory, the equations, and the computer programs, which by now could have covered enough rolls of wallpaper to decorate a small room. They discussed the problem with one another and with colleagues, listened to critiques, and incorporated suggestions that seemed apt.

They continued to feed the computer the initial conditions that represented their best guess about the satellite's orbit. Then the computer would run through the calculations to produce the Doppler shift associated with those elements. The process was iterative, with small changes being made to the starting conditions and fresh Doppler curves generated for each set of conditions. Sometimes, it looked as though they had found as good a fit as they were going to find. Listening still to their intuition, they were convinced they could find a set of Keplerian elements that would generate a theoretical Doppler curve even closer to the experimental data. The new Keplerian elements would sometimes make the curve first grow away from a fit, but then as small changes were made, the theoretical curve would grow closer than it had been before. The process was a little like finding yourself in a valley, only to climb through trees and find a deeper valley beyond. But still, the match was not good enough.

They grew despondent, questioned themselves, and searched some more. They noticed that from one experimental curve to the next, the frequency of the transmitter varied. *Sputnik*'s oscillator was not stable. Exactly as those planning tracking for the IGY had feared, transmission frequency could not be treated as a constant. But Guier and Weiffenbach needed to know the transmitter frequency in order to generate theoretical Doppler curves.

Now they started varying the value of the transmitter frequency. The same principle applied as when varying the Keplerian elements: if they could find a fit between theory and experiment, then the theoretical values of transmitter frequency as well as of the Keplerian elements must be the values of the actual setup. The task was now way beyond what would have been possible without the recently installed Univac.

The whole process was barreling along when *Sputnik I* suddenly stopped transmitting. Guier and Weiffenbach had data to work with retrospectively, much of which still had to be reduced. In practice therefore, *Sputnik's*, silence made little difference, but psychologically it was disappointing to Guier. The pair were also beginning to suspect that an aspect of the physical world that they had thought could be ignored—the ionosphere—was, in fact, significant. They were right. But to solve the problem they needed another satellite transmitting two frequencies.

Once again the Soviets obliged. *Sputnik II* was launched November 3, 1957. It transmitted at twenty and forty megahertz. With these two frequencies they could make the correction necessary to solve their problem. Theory and experiment started to come together.

Shortly after the launch of *Sputnik II,* they did it—found a theoretical curve that gave them *Sputnik's* orbit. It was nowhere near as accurate as later orbital determinations would be, but they had a method to determine all the Keplerian elements and the transmitter frequency from data collected during one satellite pass over one ground station. Guier and Weiffenbach were delighted. It was the first time, recalls Guier, that they jumped for joy. Henry Elliott, who with Harry Zinc had joined Guier and Weiffenbach shortly after their work started, recalls being ecstatic. Bitterli, who went on to code far more complicated programs than a least-squares algorithm, came to see this work as one of the most important accomplishments of his professional career.

Guier and Weiffenbach were now on the edge of new ground, surveying terrain that was full of quagmires and briars. The basic computational and statistical methods existed for a novel method of orbital determination and prediction, one that did not rely, as did every other, on measurements of angle. The same computational and statistical techniques would be applied, in an inverted manner, to satellite navigation.

Guier and Weiffenbach suspected that their method was correct to within two or three miles. McClure suggested checking their predictions against orbits determined by other groups. They compared their results with those from Jodrell Bank and the Royal Aircraft Establishment and found good agreement. Later they checked their orbital predictions for *Sputnik* II and the first American satellite, *Explorer I,* with Minitrack's results. Again, the methods gave comparable accuracies.

The insight that each orbital trajectory had a unique Doppler shift associated with it was not new physics; rather, it was an inevitable consequence of the Doppler phenomenon in the dynamic, three-dimensional world. Nevertheless, Guier and Weiffenbach were the first to recognize that consequence and its implications for orbital determination.

More importantly, they got the process to work in the real world, where very little behaves ideally. If they had stopped to think about the task too much, they might have concluded that the method would never work to any useful degree of accuracy. If they had known more than they did initially about orbital mechanics, the ionosphere, or the discussions at the IGY, they might not have started their work as they did and thus not developed their techniques.

Some people who heard of their research found it hard to believe that all of the Keplerian elements and the transmitter frequency could be found by fitting a theoretical curve to one experimental set of data. But the trajectory of every satellite pass, part of a definable orbit, has a unique Doppler shift. And in three dimensions the ambiguities of two dimensions are eliminated. The method, to use their terminology, worked because of its exquisite sensitivity to the range rate. It was an excellent example of applied physics, and the danger is that it might sound trivial in the telling, particularly a telling that leaves out the mathematics. To fall back on scientific cliché, the task was nontrivial.

From *Sputnik II* to Transit

On November 8, 1957, Ralph Gibson dipped into the director's discretionary fund for $20,000 to fund Project D-54—to determine a
satellite orbit from Doppler data, and he assigned technical and engineering support to Guier and Weiffenbach.

Shortly afterwards, Guier and Weiffenbach briefed their colleagues in
the research center, some of whom would later play an important part in
the development of Transit. One, Harold Black, was bowled over. Guier
had explained methods and coordinate systems that were on the edge of
his understanding, and Black knew that he wanted to be a part of the
work. He would be, but not for a few more months. Not until the Transit
program was underway.

In the closing months of 1957, Guier and Weiffenbach were tackling
the problem of the ionosphere—the region of the earth's upper atmosphere where solar energy separates electrons from atoms and molecules,
creating layers of free electrons that reflect or refract radio waves. Reflected
waves pass round the curve of the Earth and carry transmissions from, say,
Voice of America, or provide medium- and high-frequency channels for
voice communication. Higher frequencies pass through the ionosphere,
enabling communication between the ground and a satellite.

On passage through the ionosphere, radio waves interact with the
free electrons, and the frequency of a signal received on Earth from a satellite appears different from the frequency transmitted by the satellite. This
muddies the water if you are trying to relate the Doppler shift to the satellite's motion. There is a qualitative explanation that gives a rough idea of
what is happening.

Radio waves are, of course, examples of electromagnetic radiation.
Sputnik's oscillator was creating an electric field (with its associated mag-

netic field), and the influence of the field extended through space. The field's influence set the free electrons of the ionosphere oscillating with the same frequency as the field. The oscillating electrons set up their own electromagnetic fields, which, in turn, extended their influence out through space. The fields from *Sputnik*'s oscillator and from the free electrons were of the same frequency but were not exactly in step. They overlapped. If one returns to the wave analogy of electromagnetic radiation, it as though the crests of the waves from the field generated by the free electrons occurred at a slightly different time than the crests of the waves from the satellite oscillator's field. So a receiver on Earth detected more wave crests over a given time than it would if the ionosphere were to conveniently disappear (convenient at least for this application).

The Doppler shift (received minus transmitted frequency) is related to the range rate, that is, to the way the satellite's position with respect to the lab was changing. So when the received frequency was altered by the presence of the ionosphere, the satellite's path appeared to be different from what it actually was.

How could the receiver distinguish between the number of crests received in a given time as a result of ionospheric refraction as distinct from the number of crests detected because of the Doppler shift?

It couldn't. And because *Sputnik I* generated only one frequency, there wasn't much that Guier and Weiffenbach could then do by judiciously juggling theory and observation. Fortunately, the Soviets launched *Sputnik II,* and Weiffenbach started recording both frequencies. By comparing the two signals received, and knowing, from theory, that the ionospheric effect was roughly inversely proportional to the square of the transmitter frequency, they were able to calculate the amount by which the ionosphere altered the received frequency. They were able, therefore, to remove the effects of the ionosphere from their experimental data. Once Guier and Weiffenbach corrected for ionospheric refraction, the frequencies computed were those due to the satellite's motion and not to passage through the ionosphere. Needless to say, this again sounds far simpler than it was, and considerable effort subsequently went into theoretical and experimental explorations of the ionosphere's nature.

Fortunately, the IGY was generating new observations of the ionosphere. Of particular importance for greater theoretical understanding were the Naval Research Laboratory's data showing that the then existing simplified description of the ionosphere was less complete than had been thought previously. The nonuniform electron density through a cross section of the ionosphere raised concerns at APL that more than two frequencies would be needed to correct for ionospheric refraction with sufficient accuracy. Guier's and Weiffenbach's work with *Sputnik II* did not settle this question.

Another potential complication was that electron density, which influences the amount of refraction, varies with the amount of energy the sun is pouring onto the earth. There are, for example, more free electrons at noon than at night, and the density varies according to the latitude, the time of year, and the stage of the solar cycle.

Thus in November 1957, the ionosphere was the first quagmire that Guier and Weiffenbach encountered as they analyzed their Doppler data. Their experience led them to conclude in the first Transit proposal, erroneously as it turned out, that the ionosphere would be the biggest obstacle to developing a navigation system.

Once the Transit program was underway, Weiffenbach and then Guier and then others tackled the ionospheric physics in greater detail and with more sophistication than in Guier and Weiffenbach's initial paper in April 1958, entitled *Theoretical Analysis of the Doppler Radio Signals from Earth Satellites.*

Weiffenbach went back to basics, studying Maxwell's equations describing the behavior of electromagnetic waves as well as the equations that provide a theoretical and simplified description of the ionosphere. He collected much of the published data and satisfied himself that theory and experiment seemed to be in synch. Weiffenbach's aim was to find the optimum frequency for Transit given the constraints that physics and current or foreseeable technology imposed. Ideally, the highest possible value should be chosen, but electronic components of the day did not work at the higher frequencies. Such juggling of theory and practicality was an essential aspect of determining the early specifications for Transit, and it focused the minds of many other satellite designers.

By the end of 1958, Weiffenbach had concluded that two frequencies should do the job. The analysis he submitted to Richard Kershner, the team leader, warned that only experiment would clarify the situation, and the first experimental Transit satellite was aimed at doing just that.

Guier's later basic theoretical analysis, bolstered by results from early experimental satellites and discussions with Weiffenbach, showed that for Transit's purposes, physical phenomena such as the irregularity of electron density could be ignored, and that two frequencies provided an accurate enough correction for ionospheric refraction.

Shortly after the first Transit proposal was written in April 1958, they realized that the earth's magnetic field would also affect passage of radio waves through the ionosphere. Therefore, Weiffenbach, who became responsible for space physics and instrumentation at APL, recommended circularly polarized transmitters. Such transmitters were not unusual, but Henry Riblet, who went on to head space physics and instrumentation at the lab, had to modify existing designs so that they would be suitable for the spherical surface of Transit. Having chosen circular polarization, Transit's designers were constrained in their approach to stabilizing the satellite in orbit. Thus the physics affects the technology and each technological decision affects others.

As 1957 drew to a close, Guier, later responsible for space analysis and computation, was refining the mathematical model. Weiffenbach was gathering data and improving the experimental setup, and together they continued to explore what the model could tell them about orbits. In the meantime they were reading madly, educating themselves in ionospheric physics and orbital mechanics.

Around this time they met someone who was to become their "good angel." His name was John O'Keefe, and he was working with the Vanguard tracking team, was in fact at the radio tracking station near Washington D.C. on the morning after the launch of *Sputnik I*.

O'Keefe, who was to become a pioneer of satellite geodesy, had studied the moon's motion in search of clues to the nature of the earth's gravitational field and had eagerly anticipated the launch of satellites. For the previous year, in preparation for the space age, he had studied orbital mechanics every morning before going to work. He was devastated when the Soviets launched first.

O'Keefe, whose son heard *Sputnik*'s Doppler shift on a ham radio set, was intrigued when he learned of the quality of Guier's and Weiffenbach's Doppler data. He went to APL to take a look and was impressed. When Guier and Weiffenbach determined *Sputnik*'s orbit, O'Keefe was astounded; he didn't think they would be able to get that much information from the data.

Yet there was actually even more in the data. Later, the same curve-fitting technique enabled them to vary values for ionospheric refraction and thus to find indices of refraction in addition to transmitter frequency and the orbital parameters.

O'Keefe tweaked Guier and Weiffenbach about their lack of knowledge of the ionosphere and pointed out, too, that the Earth's gravitational field would turn out to be far more complicated than people yet realized. He brought much humor and knowledge of physics to their relationship.

As a regular attendee at IGY meetings, O'Keefe was up-to-date on emerging results. In the following year, he would analyze observations of *Vanguard 1,* known as "the grapefruit," and he recognized that for the satellite to be in the observed orbit, the Earth must be pear-shaped in addition to being oblate. The southern hemisphere, in effect, is larger than the northern hemisphere.

The pear-shaped term for the Earth eventually became the second brush stroke on the canvas of that very simple mathematical model that Guier had initially constructed for his and Weiffenbach's research during the autumn of 1957. And when the gravitational consequences of the pear-shaped earth were folded into the model, the fit between theoretical and experimental curves improved yet again, reducing the error of the orbital determination and prediction.

Immediately after the launch of *Sputnik I,* O'Keefe had very little data to work with. Hence his fascination with Guier's and Weiffenbach's Doppler curves. He invited them "downtown" to Washington D.C. to meet the Vanguard's Minitrack team. At the time the Minitrack group was inundated. Calls were coming in from many people whom they really didn't want to talk to. Guier and Weiffenbach fell into the category of nuisance, and to their chagrin, the Minitrack people dismissed them and their work. The rebuff rankled. Weiffenbach recalls thinking, "OK, we'll show you; you will take notice."

Throughout the early part of 1958, Guier and Weiffenbach focused on improving orbital determination. By early spring, they thought they had the satellite's position to within two or three miles, and they began writing their key paper on the theoretical analysis of Doppler data. But what next?

It was their boss, Frank "Mack" McClure, head of the lab's research center, who took the next step. McClure seems to have been a complicated character. Various descriptions crop up from those who knew him:

smart as hell; prone to terrific rows; blunt; impatient; enormous ego (I doubt this was a unique characteristic); someone who "didn't fool himself about himself" recalls Weiffenbach. Certainly he seems to have impressed the research center's scientists with the breadth and depth of his knowledge of physics.

At this time McClure's main interest was in solving the problem of instabilities in the burning of fuel in solid-fuel rockets, one of the critical technologies for the Polaris missiles. Hence he was spending a lot of time at Special Projects, as was the man who became the Transit team leader—Richard Kershner.

On Monday, March 17, 1958, Frank McClure called Guier and Weiffenbach to his office. He told them to close the door; a clue, they knew, that something interesting or classified would be said. McClure asked them, Can you really do what you say you can? In the manner of the best courtroom attorneys, McClure was asking a question to which he was sure he already knew the answer. Guier and Weiffenbach did not know this, but they answered yes.

In that case, McClure told them, if you can determine a satellite's orbit by analyzing Doppler data received at a known position on the earth, it should be possible to determine position at sea by receiving Doppler data from a satellite in a known orbit. The navigation problem, in fact, would be easier, argued McClure, because only two values—latitude and longitude—would have to be determined.

McClure sketched out the concept. The satellite would transmit two continuous waves, which would allow a receiving station on a submarine to record two Doppler curves and to correct for ionospheric effects. The satellite would also broadcast its position at the time of the transmission.

The submarine's inertial guidance system could then provide an estimated value of latitude and longitude. A computer program would generate the Doppler curves that a submarine at that estimated location would receive from a satellite at a known position. It would then continue to generate Doppler curves that would be received at nearby values of latitude and longitude. When a best fit was found between the estimated and received Doppler curves, those values of latitude and longitude would be the latitude and longitude of the submarine.

Thus the method that Guier and Weiffenbach had developed would be used in two ways. First, to find a satellite's orbit by observing it from ground stations at well-determined locations, allowing a prediction of its position. The positions for the next twelve hours would then be uploaded

From left to right: William H. Guier, Frank T. McClure and George C. Weiffenbach. Drs. Guier and Weiffenbach first employed Doppler tracking to determine the orbits of satellites when tracking *Sputnik*. Dr. McClure used this principle as a basis for inventing a worldwide navigation system. Courtesy of Applied Physics Laboratory.

twice a day. Second, the same computational and statistical techniques would provide a means of fixing latitude and longitude.

McClure asked them what they thought. They agreed that it sounded plausible. Go away, McClure told them, and do an error analysis.

They did. After the computer simulations they had been working with, simulating the errors in the navigation scheme proposed by McClure was, they say, next to trivial. They looked at the answers and didn't believe them. They seemed too good. They wondered whether they had left something out, but they knew that they had not. So they increased the errors in the assumed depth and speeds of the submarines as well as several other parameters. Then they went back and told McClure that it would work, and work beautifully.

McClure smiled and said, "I know."

McClure told Guier that he had had the idea the weekend before and had called Richard Kershner. The two of them, who frequently spent time together away from the lab, had fleshed out the idea over the weekend. They had concluded that the navigation method would work if Guier and Weiffenbach could do what they said they could.

On the day after his meeting with Guier and Weiffenbach, McClure sent a memo to Gibson, APL's director. The topic was "satellite Doppler as a means of ship navigation." McClure suggested that APL's patent group should consider seeking protection for the idea. He pointed out that "While the possible importance of this system to the Polaris weapons system is clear ... an extension of thinking in regard to peacetime use is to me quite exciting." Other advantages did not escape him, and he added that the establishment of such a system " ... might provide a wonderful cold war opportunity.... This would put the U.S. in the position of being the first nation to offer worldwide service through its venture into outer space."

Work on a technical proposal began immediately.

Developing specifications for a complex engineering system is like planning an elaborate meal without reference to recipes, knowing only what combinations are most likely to work, what fresh produce is likely to be available, and making last-minute adjustments for what is not. The occasion for the meal guides the process. First there is the overall plan of suitable and compatible courses. Then comes the detail, working out all of the recipes, testing quantities, temperatures, cooking times.

McClure defined the occasion and made clear what the "diners" would swallow. He knew the technical needs of Polaris in detail, but of equal importance was his sense of the technological likes and dislikes of Special Projects. If he thought Special Projects wouldn't like a particular technical approach, he would point that out.

Several factors, therefore, converged to ensure that the Transit proposal would win the backing that it did in the face of opposing schemes that were emerging for satellite-based navigation. First, APL was presenting a good idea and had some smart people able to see it through from theoretical analysis to engineering application. Second, the lab had an acknowledged history of completing complex and difficult applied physics and engineering projects, starting with the proximity fuse. (The Naval Research Laboratory's experience with Vanguard after the success of Viking, however, showed that previous success is not always a harbinger of future triumphs.) Neither factor would have been enough alone, but Transit would clearly have been dead in the water without the lab's expertise and reputation. Additionally, of the emerging ideas for satellite navigation, Transit met the needs of Special Projects and, because of the Brickbat-01 status of Polaris, was compatible with national priorities.

Of the competing schemes, one, taking angles to emitting radio waves was impossible because the antennas needed would have been far too large for submarines. Even had the antennas been usable by the surface fleet, the scheme would not have been serious competition for Transit because the surface fleet was not expressing a strong desire for improved position fixing, nor had its missions the same high priority as those of Special Projects.

Another scheme, radar ranging, might have met the needs of the U.S. Air Force, but although Air Force missions did have a high priority and the Air Force had proven in its ten years of existence to be highly successful at lobbying the government, the preferred scheme within the Department of Defense at that time was to develop inertial guidance for aircraft. Since aircraft are not aloft for months at a time, the errors do not have as long to accumulate as they do during a submarine's tour of duty. (Radar ranging resurfaced as the technical basis for the GPS satellites, which have replaced Transit.)

Also operating in APL's favor was the lab's relationship, as the consultant assessing the Polaris weapons system overall, with Special Projects. The organizational relationship was bolstered by professional relationships at all levels and by McClure's political astuteness in making his inside knowledge of Special Projects work to further the technical acceptability of the Transit proposal. In short, the people who proposed Transit were the right people, in the right place, at the right time, and with the political and scientific smarts to see it through.

There was another influential factor, and that was Richard Kershner's reputation. Kershner's approach to technical problems was similar to that taken by Captain Levering Smith, then technical director of Special Projects. Commander "Chuck" Pollow, who assisted in the day-to-day management of Transit, is convinced that the perception that Special Projects and Smith had of Kershner was critical to Transit's initial acceptance. And the Transit team members are convinced that Kershner was critical to the success of Transit.

Kershner's Roulette

9

You have to give yourself a chance to get lucky.
There's plenty of time, if we work hard enough.

—Richard Kershner, Transit project leader

If people leave a legacy in the eyes and voices of those who knew them well, then Richard Kershner should rest easy. Kershner led the Transit team with an authority bestowed willingly by those who worked for him.

He instituted a policy of "cradle to grave" engineering, which means that the person who designs a particular component also has responsibility for testing it. This was not (and is not) a common practice, but Transit team members perceive it as having been critical to the project's success. Kershner eschewed line management as much as possible. If he had a question about a particular component, he would go directly to the person building that component rather than the department head. The policy was applied widely. Someone working for Guier, for example, could and would go around Guier if he thought an issue would be discussed more profitably with someone else.

There were the inevitable personality conflicts and disagreements. Yet a widely held memory among the Transit team is that the optimum technical solution determined the outcome of an argument, not an individual's position in the hierarchy. All recall that Kershner was the final arbiter. He would tell them, "you together have n votes, and I have n plus one votes." Yet his decisions seem to have been accepted because of his character, not because of the authority vested in his position. His most admired characteristic was his acceptance of responsibility when things went wrong, and over the years many things did go wrong in the Transit program. Kershner would say that whoever did not like what had happened would have to deal with him first. More than thirty-five years after the project started and fourteen years after Richard Kershner's death, the Transit team's respect is still palpable.

One can look for the reasons among articles Kershner wrote about management practice. They are as dry and irrelevant as, I suspect, most formulas for successful management. Perhaps they could not be applied by someone other than himself; he seems to have had a natural tact and respect for his staff that won him great loyalty.

Immediately after McClure had explained his idea for satellite navigation, Kershner was brought in to help put together a proposal. He was joined by Guier, Weiffenbach, and Bob Newton, a reticent, elegant Anglophile. During Transit's development, Newton was to contribute to nearly all areas of the physics. In future years, these four would commemorate their early collaboration by celebrating the anniversary of the first successful Transit launch.

Their initial proposal was completed by April 4, 1958. It was marked "confidential," a low level of classification. Submarines were mentioned only in passing, even though Polaris provided the impetus for the proposal. Informal discussions about navigation for Polaris must have been going on at high levels of APL and the Pentagon.

During the next few years, details of the proposal would change. The ionosphere, for example, deemed at this time to be the biggest problem, was soon supplanted in importance by the need to understand the earth's gravitational field. Nevertheless, a surprising amount of the original proposal was to remain.

For the first draft, Guier tackled computing, Newton dealt with the satellite's motion, and Weiffenbach took on the ionosphere and assessment of error—frequency drift, for example. Kershner worked on the overall system design. After the first draft, they all worked on all sections, critiquing and revising, guided by McClure's intimate knowledge of what Special Projects wanted and would accept.

A worldwide "satellite Doppler navigation system" for all weather was possible, argued the proposal, because of the recent developments showing that it was technically feasible to establish artificial satellites in predetermined orbits with lifetimes measured in years. A position fix with a CEP (circular error probable—a circle of radius within which there is a 50% probability of finding an object) of half a nautical mile should be possible immediately. CEPs of one-tenth of a nautical mile, about six hundred feet, were likely in the near future, and such accuracies, stated the proposal, were greater than those of any known military requirements. Transit team members say that the first operational satellite launched at the end of 1963 achieved a CEP of one-tenth of a mile. Even though there were still problems to be solved, this was an impressive feat.

An alternative approach to navigation, involving further development of a coastal radio ranging system called Loran, would also have been accurate, but only when the submarines were within range of its signal. Loran, says a former British liaison officer to Special Projects, would have been too limiting for Polaris.

A particular strength, the proposal pointed out, was the passive nature of the proposed navigation system, which would not betray a submarine's position. The submarine would need only to approach the surface and to deploy its antenna at night for about eight minutes.

Various approaches to calculating latitude and longitude were outlined. One involved our old friend "miss distance," that is, the range at closest approach. Given the miss distance, one could plot a line on a map representing positions for which the submarine might be at the corresponding range from the satellite. If this process were repeated for signals from two satellites, the latitude and longitude would be the place where the lines of position intersected. Conceptually, this was an approach familiar to navigators.

APL, of course, was not trying to persuade anyone that relying on miss distance was a good approach, and the proposal went on to describe in general terms the computational and statistical techniques developed by Guier and Weiffenbach during their research into orbital determination and to say that these would be applied for position fixing. The greater accuracy attainable by comparing the curves of the theoretically generated and received Doppler signals at many points was pointed out. They needed to make this argument explicitly because many opponents at the time, recalls Guier, though that the method was "out in left field."

To recap, the Transit proposal was essentially this: A satellite in a known position would broadcast two continuous waves. A comparison of the two signals would allow the submarine's computer to eliminate the effects of ionospheric refraction. That same computer would then generate—for different latitudes and longitudes—the Doppler curves that the submarine would receive from a satellite in that position. The inertial navigation unit would provide an initial estimate of position to the computer, which would run through a number of iterations until it found the latitude and longitude that generated the Doppler curve that best fitted that received from the satellite. These values of latitude and longitude would then be fed to the inertial navigation system to automatically correct its output.

The system, which was yet to be named Transit, was explained through the example of a satellite in a polar orbit at an altitude of four hundred miles. Such a satellite takes ninety-six minutes to complete one orbit. During these ninety-six minutes, the earth turns through about twenty-four degrees of longitude. The point immediately below the satellite—the sub-satellite point—moves westward by an equivalent number of miles, depending on the latitude. Thus, how frequently one could "see" such a satellite during the course of a day depends on latitude. Near the poles, the satellite would be visible on every orbit; near the equator, less often.

Simply seeing the satellite, as Guier and Weiffenbach had come to understand by observing *Explorer I,* would not be enough. The relationship of the orbit to receiving equipment on the ground had to fall within certain limits. If the arc of the orbit were too distant, the recorded Doppler shift would change far more gradually with time than if the satellite were nearer. The quantity of positional information would then be less per unit time, and with constant noise being received, the so-called signal-to-noise ratio would be too low for optimum accuracy. Moreover, the signal would be passing through a greater distance and would be subject to more atmospheric refraction, which would obscure the more gradual Doppler changes of the distant satellite. On the other hand, if the satellite were to pass nearly overhead, the computations would result in an error in longitude. In between, there was a wide region in which the total navigational error would be acceptably small and would be insensitive to the geometry of the satellite pass.

These limitations were pointed out, but the proposal did not recommend a particular number of satellites or an orbital configuration. Nevertheless, one begins to see how the laws of physics and the requirements of the job to be done, for example worldwide navigation versus position fixing in the arctic, combine to impose limits on the configuration of orbital inclinations and altitudes.

A vital part of the system would be the oscillators producing the continuous waves for the Doppler measurements and the reference frequency in the submarine's receiver. The stability of the oscillators and the error in the frequency measurement defined the system's theoretical limit of accuracy—the limit achievable if all other problems were solved.

The first proposal asserted that the state of the art for oscillators meant that a CEP of a thirtieth of a nautical mile (roughly two hundred feet) was theoretically possible. Although the proposal had made clear that this accuracy was dependent on a number of assumptions, including a favorable geometry between satellite and receiving station, critics thought that APL was saying that it could achieve this number regularly in practice. When the figure leaked out, McClure had to field outraged and disbelieving phone calls, including one from Bill Markowitz, head of the Naval Observatory. Eventually, Transit surpassed even this accuracy, but in 1958 the proposal's claims seemed implausible.

The situation was not helped by the fact that many of the critics either did not understand or did not want to understand the computational and statistical approach APL was proposing. Guier and Weiffenbach recall one group that converted their data to miss distances, calculated, inevitably, a very poor orbit, and so dismissed their work.

Transit's history was similar to that of many technological developments. First, Guier and Weiffenbach explored something new and unexpected, initially with no clear idea in mind. Then they found a purpose for their research—orbital determination. In March, McClure entered the stage with a way of exploiting their research for a practical application. McClure brought with him Richard Kershner, someone who today would be called a systems engineer. Kershner supervised the development of a proposal, which, because it served a military need, won funding. It encountered opposition and was not widely accepted for some time.

One could write of this last aspect of Transit's development, that there is "nothing new under the sun," because the satellite's development has much in common with the pursuit of longitude, the critical navigational problem that occupied physicists of the seventeenth and eighteenth centuries—some of these same physicists who laid the scientific foundations for Transit.

A diversion into history is too hard to resist. Geopolitical tensions between two superpowers—Spain and Portugal—provided the initial impetus for the pursuit of longitude. The two countries had a territorial dispute that could not be resolved because they did not know the location of the meridian in the Atlantic ocean that separated their claims to sovereignty. Before a solution was found, there were political hearings, testimony from distinguished scientists, competing technical solutions, pleas for public funding, and suspicion of the new technology from naval officers. Once a solution to the longitude problem was found, it became invaluable to commercial shipping interests.

In the seventeenth century, the problem of determining longitude came down to the need for a means of telling the time accurately at sea. If a clock of some kind were set according to time at a reference longitude, say Greenwich, and local noon occurred at 2 P.M. Greenwich time, then you knew you were thirty degrees west of Greenwich. (The earth turns through fifteen degrees of longitude per hour).

Telling the time at sea was acknowledged to be very difficult. On land there was no problem. Pendulum clocks were accurate to within a few seconds, but they did not work well on the pitching and rolling deck of a ship. First the Portuguese, then the Spanish, the Dutch, and the French offered rewards for ideas leading to a method to tell the time at sea. Every crank in Europe responded. Overwhelmed by dubious proposals, Spanish officials returned a promising idea from an Italian named Galileo Galilei that suggested taking advantage of the times of eclipse and emergence of the moons of Jupiter. Given the difficulties of stabilizing a pendulum at sea, the idea was good. But it languished in bureaucracy.

A century after the Portuguese first tackled the problem, the English Parliament bestirred itself in an effort to placate a public scandalized when, in 1707, Admiral Sir Clowdisley Shovell led the fleet onto rocks in bad weather off the tip of Cornwall. Some two thousand men died. Every navigator but one had agreed that the fleet was southeast of its actual position just off the coast of Brittany.

After more losses of men and goods, a petition signed by several captains of Her Majesty's ships, merchants of London and commanders of merchantmen was set to the House of Commons in 1714. Though couched in florid, formal language, its message was clear: Parliament should act. The petitioners suggested that public money be offered as an incentive to anyone able to invent a method of determining longitude at sea.

Parliament sought expert advice, calling, among others, Isaac Newton and Edmund Halley to testify before the parliamentary committee of the House of Commons on June 11, 1714. Newton was not encouraging. He spoke of " . . . several projects, true in practice, but difficult to execute."

Rather than celestial schemes, he favored the development of an accurate clock, telling members that of the various options for determining longitude, "One is by watch to keep time exactly, but by reason of the motion of the ship at sea, the variations of heat and cold, wet and dry and the difference in gravity in different latitudes, such a watch has not yet been made." He offered no engineering advice.

A few weeks after this testimony, Parliament enacted legislation offering a reward to the first inventor to develop a clock that met specifications laid out in the act. The award was £20,000, which for the time was a considerable amount of money. The act also established a Board of Longitude, which was given the task of awarding smaller grants for promising ideas as well as responsibility for determining when the terms of the act had been met. An unschooled but skilled cabinetmaker called John Harrison received several of these grants and eventually developed a chronometer (known among the cognoscenti as "chronometer No 4") that fulfilled the act's specifications.

The board proved reluctant to pay the full £20,000. There were conflicts of interest, ownership disputes, and numerous sea trials. Undeterred, the Royal Society honored Harrison, but the monetary reward was still not forthcoming. Then the board paid up in part. Unsatisfied, Harrison appealed to the king, who offered to appear under a lesser title to argue the case before the Commons. Eventually, the board paid out the full amount.

Another time, another place: In the late spring of 1958, Kershner prepared to do battle for what would be a controversial new approach to navigation. Gibson, McClure, and Milton Eisenhower, the president of Johns Hopkins University and brother to the other president, were playing their parts too.

Kershner made several trips to the Pentagon, often taking Guier and Weiffenbach with him. At first, Guier and Weiffenbach felt apprehensive and out of their depth. Amidst these efforts, Weiffenbach was awarded his Ph.D., something of a sidebar as APL fought competing navigation proposals and criticism. The Jet Propulsion Laboratory, which had a deservedly high reputation, posed a particular threat when it challenged APL's error analysis. Eventually, there came a meeting at which Guier and Weiffenbach argued for an opportunity to fly an oscillator on someone else's satellite. This would give them a chance to prove the concept and to show that an ionospheric correction was possible. Their audience, though, had already been convinced of the validity of APL's approach to developing a navigation satellite, and Guier and Weiffenbach left with the promise of funding for a satellite, not just an oscillator. (An oscillator was flown on the Department of Defense's *Discover* satellites as part of the effort to determine the earth's gravitational field). This happened sometime in the summer of 1958, and APL had by now developed technical plans for eight experimental satellites. Their aim was to have a prototype operational satellite aloft by the end of 1962.

Others had joined the project since April. In July, Weiffenbach sent John Hamblen the operating characteristics for the satellite's antenna. Weiffenbach was also designing the first oscillator and reviewing what was known about the ionosphere. Harold Black had his wish and was working for Guier and Newton on the computational effort for orbital determination and prediction. Joy Hook, a skilled programmer, had joined the team and was rapidly becoming invaluable. Many subroutines were being completed, including those for organizing the recorded data on magnetic tape, for applying the refraction correction, and for curve fitting. There was, says Black, a quiet desperation about much of the software effort. Very few people understood it well.

In November 1958, three of APL's engineers visited Iowa State University, seeking advice from George Ludwig in James Van Allen's group about the kind of environment their satellite would encounter in space. They returned with information about how ISU built and tested its satellites and about temperature control. They picked up schematics for the *Explorer III* satellite, which had been launched on March 26, as well as information about the satellite's radiation observations. These observations, which were the first data supporting the hypothesized existence of radiation belts, had not yet been widely circulated. They were to be important to Transit because the satellites would be in orbits that took them through the newly discovered radiation belts.

As work on Transit progressed, more detailed questions emerged. How many transistors, for example, would be needed in the clock circuit or the oscillator circuit? The original proposal of April 4 had said that the satellite would be fully transistorized, so these questions were significant, especially at a time when a transistor could cost as much as ninety dollars. How much power would the circuits need? How would they be packed and into what volume? The original specifications were for 293 transistors

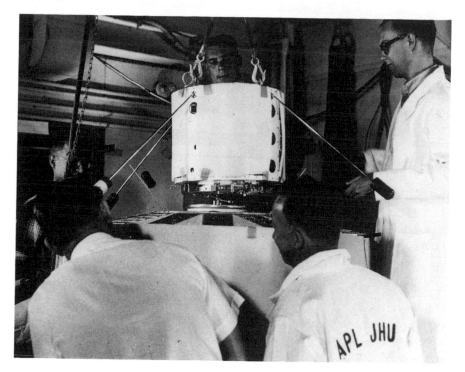

Researchers from the Johns Hopkins University Applied Physics Laboratory position a satellite from Iowa State University on top of the Laboratory's Transit 4A satellite in June 1961. Courtesy of Applied Physics Laboratory.

or diodes. With an estimated packing volume of two cubic inches per transistor or diode, the electronics could not be packed into a volume of less than 586 cubic inches and would weigh eighteen pounds. During the rest of 1958 and through 1959, these numbers were refined.

Another difficulty arose—how to affix the electronics to the circuit boards. After some faulty starts with soldering and some burned-out transistors, the engineers decided to weld the components. But different coefficients of expansion between the connections and the components led to many breakdowns. Reliability did not really improve until the advent of integrated circuits in the mid 1960s.

On December 15, in the midst of these ongoing problems, the Advanced Research Projects Agency (ARPA) awarded APL full funding of $1,023,000 for the Transit project. Although ARPA awarded the contract, the work was sponsored jointly by the Special Projects Office and was administered by the Navy's Bureau of Ordnance.

The aim was for Transit to be operational by the time the Polaris submarines were on station. This goal was not met. Two submarines were on patrol by the end of 1960, three full years before Transit was operational. But the Polaris deployment had moved ahead of schedule, and the Transit development slipped by a year. Even so, Transit's development went ahead at a breakneck pace.

The objectives initially fell under two headings: the Transit program and *NAV 1*—the satellite. The Transit program was defined as the development of a shipboard navigation system, work on the earth's gravitational field, and a feasibility study of a Doppler navigation system for aircraft.

The first aim is obvious. The second became critical. The last aim, however, was curious for two reasons. First, single-satellite Doppler navigation was never going to work for aircraft because they need to determine altitude as well as latitude and longitude and thus would need to record either two passes of the same satellite or one pass each of two different satellites. For aircraft moving at, say, eight miles a minute, this was not an option, and the higher echelons of the Pentagon had decided to develop inertial navigation for aircraft. However, there was a new requirement for all-weather navigation for the aircraft of the radar picket fence, which stayed airborne for eighteen hours at a stretch. The thinking was that Transit coupled with interpolation from inertial navigation units might serve these aircraft.

The second part of the contract—*Nav-I*—comprised three satellites, which were to be delivered to Cape Canaveral in time for a launch scheduled for August 24, 1959. Once the first satellite was in orbit, its nomenclature would change from *Nav-1* to *Transit IA*. *NAV-1*'s objectives were to demonstrate the payload (another word for satellite instrumentation), the tracking stations and data processing, and the Doppler tracking methods, as well as to collect data (from analysis of Doppler curves) about the shape of the earth and its gravitational field.

The plan at the end of the 1958 was that the NAV-1 satellites would weight 270 pounds and be launched on a Thor-Able rocket toward the end of August 1959. This was considerably heavier than the original proposal's estimate of 50 pounds, but *Transit IA* was an experimental satellite. Kershner later insisted, in the face of considerable opposition from members of the Transit team, that the weight of the operational satellites be lessened to make them compatible with the less costly Scout rockets. Some of the Transit documents suggest that the aim had always been to make the oper-

An early tracking station for Transit satellites. The station located at the Johns Hopkins University Applied Physics Laboratory was for research. Courtesy of Applied Physics Laboratory.

ational satellites compatible with Scout. Team members remember that they wanted to stay with the heavier-lift launch vehicle. As was often the case, Kershner won the debate. But the transition proved more difficult than anticipated and held up the program for a year, delaying the launch of an operational navigation satellite until the end of 1963.

By February 1959, APL had made good progress with the satellite antenna. The following month, a year after McClure had called Guier and Weiffenbach into his office, seventy percent of the satellite's electronics had been built and mechanical fabrication had begun. Results of analytical work suggested that by 1963 a simplified Doppler navigation with an accuracy of one mile could be performed by hand, and that with more sophisticated techniques (curve fitting), an accuracy of one-tenth of a mile would be possible.

In the spring of 1959, the first computer program (ODP-1) for determining satellite position was nearly complete. It took the Univac 1103A twenty-four-hours to run the program in order to predict an orbit for twenty-four-hours. A year later, when the lab installed the IBM 7090 in the summer of 1960, computation time was cut to about one hour per eight hours of prediction. The 7090 was one of the world's first largely transistorized computers.

In April, two satellites underwent environmental tests in a thermal-vacuum chamber and vibration tests at the Naval Weapons Laboratory to see whether they would be able to withstand the temperature changes in space and the vibrations and accelerations of launch. Then the two satellites were placed outside on a grassy knoll while the radio frequency links and solar cells were tested.

In those early days, APL had no clean rooms to prevent contamination of the spacecraft, and Commander Pollow recalls asking someone to stop flicking cigarette ash on the satellite. The clean rooms that were then creeping into industry were not clean rooms as we know them today. They were intended, say the Transit members, to provide a psychological environment that made people more careful. APL's approach to instilling greater care relied on Kershner's cradle-to-grave policy, which was a good way of encouraging the design engineer to satisfy the test engineer.

Besides testing, April 1959 saw work begin on three permanent and two mobile tracking stations. The spring schedule called for complete installation and checkout of ground stations by June 1; the completion of thermal, vacuum, and mechanical tests on three satellites by July 1; checkout of the entire Transit tracking communication and computing system by July 15; and delivery of three completed satellites to Cape Canaveral by July 19. The scheduled launch date for the first satellite was still August 24.

By the end of July, the timetable drawn up in spring had slipped a little, but not by many days, nor is it clear from the progress reports why. According to "Kershner's roulette," a term that does not appear in the program's official progress reports but was common currency among Transit team members, any delay was supposed to be because of some other part of the complicated process of launching a satellite, never because of Transit. Whether or not the Transit team won that particular spin of Kershner's wheel, the launch date had by July been set back to September 17.

Through the summer of 1959, Guier, Newton, Black, and Hook were putting the finishing touches to the software. Kershner made informal arrangements for the Physics Research Laboratory of the Army Signal Corps to participate in Transit and to evaluate the absorptivity and emissivity of paints in an attempt to find the best materials for controlling satellite temperature in orbit. He and Gibson were in constant contact with the Cape and with the Air Force, which was responsible for the launch. Kershner was discussing tracking and orbital requirements, drawing up the checkout procedures for the launch site, and visiting candidate tracking sites. The sites were required to be free from interference from nearby sources of radio waves, and the surrounding topography had to provide clear lines of sight to the horizon, or at least to the lowest elevation from which the satellite could provide usable Doppler curves. Each site's position had to be accurately surveyed.

Finally, toward the end of July, things started to come together. There were only days to go before the satellites were to be shipped to the Cape. In the year since Guier and Weiffenbach had learned that they had funding for an experimental navigation satellite, the Transit team had worked impossibly hard. They were exhausted and exhilarated.

On July 31, the satellite in which they had most confidence was undergoing a dynamic balancing test. While the satellite was spinning at one hundred revolutions per minute, a structural support member in the test equipment failed. The satellite fell and was shattered.

Kershner spoke into the silence that followed, without anger or histrionics. He asked team members to pick up the pieces and to determine what was still in working order. Somehow, despite the sick feeling in the pits of their stomachs, Kershner's demeanor inspired them to work flat out to prepare their remaining two satellites for shipping. More than thirty years later, when Weiffenbach recalled Kershner's calmness and the team's response, there were tears in his eyes. Sooner or later, each Transit team member tells this story. It never seems very dramatic, but it clearly symbolizes why they respect Kershner.

On August 2, two, rather than three, NAV-1 satellites were delivered to the Cape.

Throughout August, the tracking stations at APL, in Texas, New Mexico, Washington State, Newfoundland, and England followed a rigorous practice schedule. APL staff went to England in the middle of the month. Negotiations were underway with AT&T to ensure that enough teletypewriter and telephone links were in place to handle communications between APL, the Cape, and the tracking sites. During the launch,

Weiffenbach found himself repeatedly reassuring an English operator that the expensive transatlantic telephone line should stay open even though the operator could hear no one speaking.

At APL's own tracking station, Henry Elliott was in place as the chief operator. During the last few weeks before the launch, Elliott encountered a number of irritants. Phase shifters arrived on September 8 but were not working. The next day, he got one working in time for a practice alert on September 10. During the day a twenty-four-hour clock arrived. On September 14, three days before launch, there was a dry run. On September 16, there was another dry run, following a timetable mimicking the next day's launch. One of the two frequency pairs that NAV-1 carried to investigate whether one pair alone would be enough to correct for ionospheric refraction was swamped by a nearby signal. If it had been the real thing, there would have been no Doppler recording of one of the frequency pairs.

September 17, 1959, was the day of the launch. A final and complete check of the station began at six in the morning. A tracking filter failed and was replaced. News came through that the launch was being held for an hour and a half. At 9:45 A.M. APL's test transmitter was switched off in preparation for the launch. An announcement was made over loudspeakers asking everyone in the lab to turn off anything that could generate a radio signal that might interfere with the satellite's signal. At last, at 10:33:27 Eastern Daylight Time, the rocket lifted off.

It was eighteen months since McClure had told Guier and Weiffenbach of his idea. In that time, the Transit team had learned how to design satellites and had designed and built facilities for mechanical, thermal, and vacuum tests, ground stations, and test equipment for their ground stations. They had carried out innumerable theoretical analyses, had written an orbital determination program, and had built and tested three satellites. They had seen their best work destroyed by a last-minute accident and had shipped two satellites to the Cape. Now the *Nav-1* satellite that would become *Transit-1A* was climbing through the atmosphere.

The Realities of
Space Exploration

10

We were pioneers, and we knew it.

—Bill Guier

P arsons auditorium was crowded. Everyone was eager to hear the news
as it was relayed from the Cape. They knew about the delays that had
accumulated during the final countdown, heard the announcement to
switch off radio frequency generators at the lab. The moments before a
launch are always tense. In the final seconds the tension was alleviated, as
the voice from the Cape intoned, "twelve, eleven, ten, eight, whoops,
seven, six, five, four, three, two, one." The Thor-Able rocket lifted off, carrying *Transit 1A* aloft. They knew that Air Force radars were tracking its
ascent; that engineers were calculating position, cross checking their slide-
rule calculations and sending course corrections to the launch vehicle as
needed. They heard the satellite's transmitters and knew that everything
was going well.

Then the transmitters stopped. For a while, no one knew what was
happening. Then came the news that the third stage had failed. In all probability, it and the satellite burned up on reentry into the atmosphere somewhere over the North Atlantic, west of Ireland. Lee DuBois, one of the
mechanical engineers, looked around the room. He saw the tears of disappointment on his colleagues' faces.

The progress reports that APL sent to ARPA were as emotionless as
those that described the shattering of their best satellite the previous
month. The launch failure, it seemed, could be ascribed to the retro rockets on the second stage. These rockets were supposed to slow the second
stage after separation of the third, so that the second stage would not interfere with the third as it coasted away prior to firing its own engine. When
the retrorockets failed, the second stage bumped into the third stage, dis-
rupting the third stage's ignition sequence.

Transit 1A's flight had lasted twenty-five minutes. Its electronics had
survived the launch. As soon as the nose fairing that protected the satellite
on liftoff had peeled away, all four frequencies were transmitted. The lab

immediately started an analysis of the telemetry, which comprised measurements of variables such as the satellite's temperatures and solar cell voltages.

APL also had some Doppler data from the short period before the signal was lost. They were incomplete; Henry Elliott's record shows that one signal was lost intermittently. For a while, an operator had locked unwittingly onto some other unknown signal. Signals from a TV station in Baltimore interfered with reception a few minutes later, and halfway through the pass, one of the tracking filters lost its lock. Nevertheless, APL learned enough to confirm "at least partially" that the ground stations' design and operation worked, according to the progress report.

With the data received the computing team also made a rough correction for ionospheric refraction. Then they set themselves a theoretical problem, imagining that the Doppler data from Transit's brief sojourn in space had, in fact, come from a satellite in orbit. They attempted their least squares fit. Though they clearly could not check the accuracy of their "orbital determination" against prediction of the satellite's position during its next orbit, they could check their results against the Air Force's radar data. They found the least squares fit was closer than it had been for the *Sputniks* and *Explorer 1* and were encouraged. Thus, though the failed launch did not yield what they had hoped for and pointed to problems that needed to be addressed, the team did learn some things.

For a definitive analysis of ionospheric corrections and to begin investigating the earth's gravitational field they needed a successful launch. The next attempt was set for April 13, 1960. By January of that year, Kershner was coordinating preparations for *Transit 1B*'s launch and that of *Transit 2A*. *Transit 1B* would be similar to the lost satellite, but *Transit 2A*, scheduled for a June launch, would test different aspects of the proposed navigation system.

Details piled on details. All over the United States, presumably in the USSR as well, teams of engineers and scientists were slowly coming to terms with the complexities of space exploration. Memos in English and Russian were written, which, if they were like Kershner's, covered an array of newly recognized problems that now are familiar to those in the space business: nose fairing insulation, loads on structures, details about an epoxy bond, maximum satellite skin temperature at launch, radio frequency links, concerns about deflection and vibration characteristics of the launch vehicle's second stage, and on and on.

Simultaneously, preparations were going forward for the satellites that would follow *1B* and *2A* in the Transit experimental series with the physics, the engineering, and computing all being developed in parallel—at a time when computing and space exploration were new.

At the ground stations, repeated preparations were made during the first three months of 1960 to track one of the Advanced Research Project Agency's Discoverer satellites, which was carrying a Transit oscillator (ToD Soc Transit on Discoverer). Transit on Discoverer was part of a program to develop precision tracking for reconnaissance satellites, and the launch was postponed repeatedly. The postponements complicated preparations for *1B*, as did expansion of the Transit control center and its communication links to encompass the other agencies that were now interested in the project and its data, including NASA and the Smithsonian Astrophysical Observatory.

During the same period, Kershner lined up the Naval Ordnance Laboratory to do magnetic measurements and experiments. The Transit team was interested in fitting its satellites with magnets to stop them from spinning (de-spin, in the industry's jargon), control their attitude, and provide stabilization.

By March, *1B* and *2A* were in the final stages of fabrication or testing. John Hamblen (who was Harry Zinc's and Henry Elliott's boss) decided that some discipline was needed. He had found out that flight hardware had been released before necessary electrical and environmental tests had been run. In a casually typed note he asked that in future those fabricating the satellite proceed to incorporate a component only if an engineer had first signed the test data sheet. Verbal assurances about a particular component, he wrote, would not do. Thus, casually, at APL and doubtless in many other labs, was the need for documentation recognized, documentation that now, assert many engineers and managers, has grown out of proportion to its usefulness.

The year advanced to Wednesday, April 13, 1960. That was a long day at APL. The launch was scheduled for 7:02 A.M. Eastern Standard Time. Once again Parson's auditorium grew crowded. Probably the room looked as it does in photographs of the launch of *Transit 4B*. The ashtray filled to overflowing on a table crowded with papers. Gibson, Kershner, and Newton formal in dark suits, others in shirt sleeves. Gibson standing, pipe in hand. Kershner in headphones, or telephone to one ear, hand covering the other. Newton seated, twisted slightly to view over his shoulder the clock held at eight minutes to launch, frowning, as was Kershner.

For *Transit 1B,* the countdown proceeded. The voice over the intercom from the Cape would have been saying things like, programmer starts ... gyros uncaged ... electrical umbilical ejects ... lift off (at 7:03 A.M.). But it was not yet time for the champagne. The satellite still had to reach orbit, which it did, though barely. Instead of the nominal 500 nautical mile circular orbit, *1B* went into an orbit with a perigee of 373 nautical miles and an apogee of 748 nautical miles. Such a result was very inaccurate by today's standards, but more precise orbits had to wait until those designing launch vehicles were able to perfect inertial guidance controls.

Transit IB's orbit was, however, sufficient to allow APL to begin work checking whether two frequencies would be adequate to correct for ionospheric refraction or whether a greater number would significantly improve the correction. The answer was that two seemed to be sufficient, though more remained to be done before this question was finally settled.

The immediate task on the first day was to determine an orbit, then to predict its position for the next twelve hours. Until midafternoon, there were computer problems. Then at 15:30 they determined their first orbit. The curves did not fit well, but they thought that this might be because the satellite was still spinning. Spinning ceased on April 19. On April 20, they determined another orbit from observations of fifteen passes at five locations. Again there was a poor fit. They decided this time that the problem was noise. Like *Transit 1A, Transit 1B* carried four frequencies for the investigation of ionospheric effects. Now they turned their attention to the second frequency pair, and the fit was better.

With the data from the second frequency pair, they determined satellite position to within 150 to 200 feet from observations of a single pass over a limited region of the earth. With data for half a day from the different tracking stations they could, assuming a simplistic model for the gravitational field and uniform air drag, determine satellite position to within one nautical mile. The longer the arc, the poorer the accuracy appeared to be. Something seemed wrong. Over and over again they looked for errors in the data and software. They could find none. It was a troubling situation.

Extrapolating from a day's observations, they then predicted the following day's orbit. This was what it was all about, developing a way of predicting an orbit so that its coordinates could be uploaded to the Transit satellites twice a day, enabling the submarines to fix position with respect to a satellite in a known position.

The Transit team looked for the satellite at the time and location they had predicted.

And then they knew they were in trouble.

There was a discrepancy of two to three miles between prediction and observation. While this much error had been acceptable when they were first establishing *Sputnik's* orbit in the fall of 1957, it was unacceptable as the basis for a navigation system. "The satellite," recalls Guier, "was all over the sky." Again, they thought that it was a problem with the programming. But it wasn't. What they had suspected but had not fully recognized, and what O'Keefe had repeatedly warned Guier and Weiffenbach about, now came to dominate the theoretical analysis of satellite motion. Earth's gravitational field was far more complicated than anyone then knew. O'Keefe, because he knew about the perturbations in the moon's orbit, was expecting that satellites in near-Earth orbits would show more pronounced perturbations, but even he could not have anticipated the huge variation and the complexity of the gravitational field that was to emerge.

For the position fixing accuracies they wanted to achieve, their knowledge of the gravitational forces perturbing near-Earth orbits needed to improve considerably.

There were precedents. Others had wrestled with apparently unruly satellites. Not least of these were the men within whose paradigms early satellite geodesists were working—Johannes Kepler and Isaac Newton. Both had struggled to understand the nature of orbits as, mystics both, they sought glimpses of fundamental truths about the universe. Kepler's focus was on the sun's satellite Mars; Newton's was on the earth's moon. In his book *The Great Mathematicians,* Henry Westren Turnbull writes, "The Moon, for instance, that refuses to go round the Earth in an exact ellipse, but has all sorts of fanciful little excursions of her own—the Moon was very trying to Isaac Newton."

And very trying would be the motion of satellites in near-Earth orbits to the early satellite geodesists who, with the technology to observe satellite motion in greater detail than could Kepler or Newton, noticed a veritable plethora of fanciful excursions. The forces causing these deviations from elliptical motion needed to be accounted for so that their effect on satellite motion could be quantified and thus orbital prediction improved. It turned out also that because the irregularities in the gravitational field are due to variations in the Earth's shape and composition, scientists reaped an unexpected and abundant scientific harvest from observations of orbits. Satellite geodesy supplied, for example, some of the evidence for the theory of continental drift and thus for theories like plate tectonics.

APL was one of the early groups observing satellite motion. They were impelled by the unlikelihood that other geodesy programs would meet Transit's needs by the time the system was scheduled to be operational, at the end of 1962.

Like other satellite geodesists around the world, the Transit team wanted to determine the "figure" of the Earth. The Earth's figure is not the topography that we see; rather it is a surface of equal gravitational potential (a geoid) that coincides with mean sea level as it would be if the sea could stretch under the continents. This geoid looks like a contour map. It has highs and lows that represent how the gravitation potential differs at a particular geographical location from what the potential would be at that point if the earth were a water-covered, radially symmetrical rotating spheroid (an ellipsoid of revolution), not subject to the gravitational pulls that cause tides. This hypothetical surface is known as the *reference ellipsoid*.

It is the differences in gravitational potential between the figure of the Earth and the reference ellipsoid that geodesists study as they seek clues to the earth's shape and structure. At first, only the deviations in motion caused by large irregularities, such as the pear-shaped Earth, were included in geoid models. Today's models include the gravitational consequences of localized irregularities in shape or density. In the mid 1990s, the most accurate geoid maps available to civilians were of what is termed "degree and order 70." Generally speaking, the higher a model is in degree and order, the more detailed is its description of gravitational potentials, in much the same way as a finer scaled topographical map gives greater detail about a piece of terrain. A geoid map, however, cannot be understood by analogy to an ordinary map. The gravitational potential at a given location is attributable not only to the local features, but also to the varying lengths of gravitational pull exerted by everything else. And the higher the degree and order of a geoid map, the more geologists can infer about the Earth's structure. Geodesists aspire in the next century to satellite-based models that will be accurate to degree and order greater than 300, the goal being to provide data that will help geophysicists to understand the earth's geological origins and history.

The road to such comprehensive understanding of our Earth opened with the launch of *Sputnik* 1. Prior to the advent of satellites, geoid maps showed modest highs and lows that were a result of local measurements of gravity. The force of gravity exerted on a satellite's motion, though, includes the sum of all the gravitational anomalies resulting from every irregularity of shape and density in the Earth. Disentangling these effects and relating them back to a specific aspect of the earth's physical nature is a little like unscrambling an egg. Nevertheless, with extensive computer modeling the job can be done.

APL produced the first American satellite geodesy map in 1960, a crude affair by comparison with those of today. Guier and Newton led this effort and found that as with orbital determination and satellite navigation, they had again provoked hostility. Their early geoid maps showed far greater highs and low than appeared in maps from presatellite days, and traditional geodesists dismissed them as amateurs.

APL continued to produce geoid maps of increasing sophistication, but much of this work was classified. Civilian scientists at places like the Smithsonian Astrophysical Observatory and the Goddard Space Flight Center soon came to dominate the field, though APL's work filtered discreetly and obliquely along some grapevines.

The lab's first gravitational model contained a value for the Earth's oblateness that was more accurate than that existing pre satellites as well as a term describing the pear-shaped Earth. Shortly afterwards Robert Newton at APL and independently the Smithsonian Astrophysical Observatory made the next big discovery, which was that the Earth is not rotationally symmetric about its axis. Just as the northern and southern hemispheres are asymmetrical, so too were the eastern and western hemispheres. A number of scientists, most particularly the Soviets, had suspected that this might be true. Later on, APL optimized their geoid maps for Transit's orbit; that is, they only unscrambled those aspects of the egg that affected polar orbits at Transit's altitude.

The principle involved in extracting information about the Earth from satellite data is simple to explain in general terms, but very difficult to apply in practice: observe the satellite, note its departure from elliptical motion—its "fanciful excursions"—and try to find (in the computer model) what aspect of the Earth's shape and structure, for example a particular dense structure or a liquid area, would give rise to the gravitational anomalies that would cause the satellite's observed departure from an ellipse.

More detailed gravitational models and ionospheric corrections enabled the orbital determination group to improve their knowledge of satellite position from between two and three kilometers in 1959 to a little under one hundred meters by the end of 1964. With problems like ionospheric refraction corrected for, other problems emerged. Would it be necessary, they wondered, to correct for refraction in the lower atmosphere? Such refraction was a source of bias in their data that could make the satellite appear to be about half a nautical mile away from its actual position. Helen Hopfield, whose dignified presence could reduce unruly Transit meetings to silence, tackled this problem, and APL made corrections for tropospheric refraction.

When the Transit group compensated for motion of the geographic poles from their mean position in the early 1970s, the satellite's position was known to within twenty-seven feet. Polar motion, caused by precession of the earth's spin axis due to the earth's nonuniform shape and structure, changes the position of a ground station by about a hundred feet per year, thus introducing a small error into the orbital determination and prediction. The error was negligible for navigators but important to surveyors.

By the time of the first launch, APL had stopped characterizing orbits solely in terms of Kepler's elements (as had other groups). First, because motion within a single orbit does not exactly obey Kepler's second law—there are small deviations, and the elements are actually average values. Second, even these average values change gradually as the orbit shifts in inertial space because of the gravitational consequences of physical irregularities in the Earth. For navigation and geodesy, averages were not good enough. It was necessary to know as exact a position as possible at given times in the orbit.

So satellite position was expressed in terms of Cartesian coordinates centered on the earth's center of mass, with one axis aligned with the earth's spin axis and the other two lying in the Earth's equatorial plane. The orbital prediction was made by finding the acceleration from Newton's second law of motion—the famous $F = ma$ that is so crucial to science and engineering, where F is force, m is mass, and a is acceleration. The value of the force acting at different parts of the orbit comes from the model of gravitation; then numerical integration of the components of acceleration, $a = F/m$, yields position and velocity at any desired instant of time.

If it had not been for the new generation of computers, typified by the IBM 7090, this work would not have been possible. The 7090 was one of the newest and best when it was installed in August 1960. It could perform 42,000 additions and subtractions per second and 5000 multiplications and divisions per second, and it could store 32,768 words (approximately 0.03 megabytes). The 7090 was almost fully transistorized, unlike the vacuum-tube Univac.

The Univac had been badly stressed by the orbital determination program, taking eight hours for eight hours worth of prediction. The IBM 7090 could do the same job in an hour. To run the early gravitation models on the Univac, which embodied only a few of the terms representing the earth's gravitational field, Guier would set aside three or four weekends. Had the Univac, which contained vacuum tubes with a mean time to failure of between 15 and 20 hours, been called upon to run the gravitational models that were to appear in the coming few years, it would undoubtedly have broken down. Even the 7090 would soon have to be updated as the gravitational model grew more intricate. Today Cray supercomputers run some of the largest models; desktop machines with Pentium or 486 chips can run models of degree and order 50.

During 1960, Newton, Guier, Black, Hook, and others prepared for the transition to the 7090. The programs had to be rewritten in an assembly language compatible with the 7090's architecture. The orbital determination program occupied four or five trays of punch cards. Woe betide the person who dropped one. And drop them they did, recalls Black, with a laugh that has an edge even after thirty years.

Black and his colleagues were also learning—painfully—about software engineering, a nascent, scarcely recognized field. Black's job was to get the orbital determination program running. He was starting with the physics developed by scientists like Newton and Guier. They generated the equations representing the physical realities, and as they understood more about what was going on they generated more equations. Black learned early to freeze the program design and fold new equations representing the physicists' deepening understanding of the situation into the orbital determination program in an orderly fashion rather than piecemeal. That, at least, was his aim; but Black's position between the scientists and the programmers who wrote and tested the code was at times unenviable. He had to force agreement out of the scientists, and he fought Guier (his immediate boss), Newton, and Kershner, telling them, "You ain't gonna change this damn thing."

In 1962, Lee Pryor, who retired in 1995 as the last project manager of Transit, arrived at the lab. Pryor had specialized in computing while taking his degree in mathematics at Pennsylvania State University. His first three months at college were spent writing programs in anticipation of the arrival of Penn State's first computer. Black says that Pryor was a godsend. When he arrived at APL, the lab was putting the finishing touches to the first "operational candidate" of the orbital determination program. "We just needed to get it out the door," recalls Pryor.

In 1962, much physics and mathematics remained to occupy the Transit scientists, but the computing was moving from their purview to that of the professional programmers like Pryor who were writing code for an operational situation rather than for research. The move was necessary because, while the scientists could write programs for their own research needs, their programs, it seems, could be cumbersome and prone to breakdown in operation.

Once work on the gravitational model was well in hand, it became apparent that the effects of air drag and the pressure of radiation from the sun would have to be considered. These were dealt with in the 1970s primarily by an elegant piece of engineering invented by Daniel De Bra from Stanford University. The navigation satellites were placed inside a second satellite. The separation between their faces was tiny. Sensors on the Transit satellite detected when the inner satellite moved toward the outer surface, and tiny rockets moved the inner satellite to compensate for these forces, before they could offset the Doppler shift. An engineering solution was necessary because the time, size, and place of the forces could not be predicted.

In the mid 1960s, the failure of solar cells threatened the reliability of the operational Transit satellites. Until this problem was solved with input from Robert Fischell, the Transit satellites tended to fail within a year of launch. Once solved, some veteran satellites exceeded twenty years in operation. The Transit team also launched the first satellite with gravity-gradient stabilization, in which an extended boom encourages the satellite to align itself with the earth's gravitational field. APL's first—unsuccessful—attempt with this technology was on a satellite known as TRAAC,

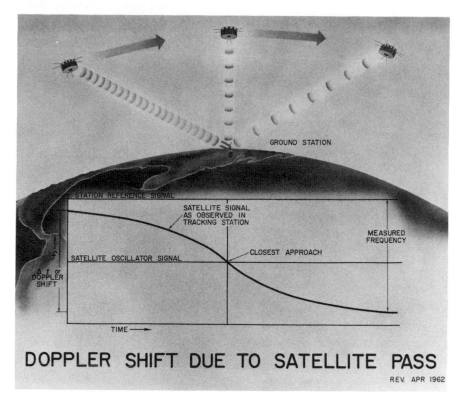

Doppler shift due to satellite pass.

which also carried instruments to explore and characterize the Van Allen radiation belts. Ironically, the satellite failed because of ionized particles created artificially by a high-altitude nuclear explosion—as did many other satellites.

TRAAC carried a poem engraved on one of the satellite's instruments. It was written by Thomas Bergen, of Yale University and is reprinted at the end of this chapter. Its mixture of hubris and wistfulness captures something of the atmosphere that surrounded the early work on satellites.

In the case of APL, that work led, of course, to the Navy's Transit Navigation Satellite System. The lab built the experimental series, the prototypes, and many of the early operational satellites. For a time, Navy Avionics built some operational satellites, but the job reverted to APL when these proved unreliable. Eventually, RCA won the commercial contract for construction. More satellites were ordered than were needed, because a problem with the solar cells that was reducing their operational life was solved after the contract was placed.

During the 1980s, under Bob Danchik's tenure as project manager, when GPS was nipping at Transit's heels, these satellites were finally launched. The last Transit satellite went into orbit in 1988.

Although there was always at least one operational Transit aloft and available for the submarines from 1964 onwards, the system was not declared fully operational until 1968. At that time four satellites provided global coverage. Not the instantaneous, precise three-dimensional position fix offered by the twenty-four-satellite constellation of GPS, but still, for the first time, an all-weather, global navigation system, a system developed initially for the military, but which evolved until ninety percent of its users were civilian.

In ten years, a newly perceived consequence of the Doppler effect in the three-dimensional world grew through all the stages necessary to design and engineer a space-based navigation system. The program began at a time when vacuum technology was giving way to transistors, when programs were written laboriously in assembly language, and when no one knew how to develop large software applications. The conditions in space were unknown. The physics of the newly entered environment had to be analyzed theoretically and understood experimentally. The complex nature of the earth's gravitational field had to be researched and a provocative new understanding of the geoid developed. Launch vehicles were imprecise in their placement of satellites, if the satellites reached orbit at all. Satellite design was a new field, with stabilization, attitude control, and communication between space and Earth all unknowns.

During the early development of Transit, the launch vehicle changed, the computers changed, and programs had to be rewritten. Ground stations and satellite test facilities were built. Programs and equipment had to be developed for the submarines. It is hardly surprising that one or two people say that they ended up in the hospital, nor that the effort is remembered vividly and affectionately. But as Pryor noted shortly before he retired in 1995, it was time for the Transit program to end.

When the Navy switched off the last Transit satellite in early 1997, it ended the longest-running singly-focused space program to date. It severed the last direct link with the opening of the space age, closing the doors on that shed on the plains of Kazakhstan and on the cold morning when Sergei Korolev thanked his exhausted and elated engineers who had launched *Sputnik I,* which Guier and Weiffenbach would track, providing the basis for Transit, which helped Polaris, America's riposte against the Soviet threat of nuclear attack, firing the rockets with whose development Korolev had been so involved, because he believed that rockets were defense and science, which they became, for both sides, as did Transit, which also became important to civilians. Here is one thread in the Cold War.

And one wonders.

What would have happened if McClure, say, or Pickering, Milton Rosen, or von Braun had met Sergei Korolev? If they had been in a room with chalk, blackboard, and a problem? Faintly, one hears the voices, discerns in imagination the energy and the imminent verbal explosions as Korolev's little finger lifts toward his eyebrow....

For a Space Prober
by Thomas G. Bergen

From time's obscure beginning, the Olympians
Have, moved by pity, anger, sometimes mirth,
Poured an abundant store of missiles down
On the resigned defenseless sons of Earth.

Hailstones and chiding thunderclaps of Jove,
Remote directives from the constellations:
Aye, the celestials have swooped down themselves,
Grim bent on miracles or incarnations.

Earth and her offspring patiently endured,
(Having no choice) and as the years rolled by
In trial and toil prepared their counterstroke—
And now tis man who dares assault the sky.

Fear not immortals, we forgive your faults,
And as we come to claim our promised place
Aim only to repay the good you gave
And warm with human love the chill of space.

Move Over, *Sputnik*

It was pretty tense because we knew that everybody was watching us, not only this country, but really the whole world, because here the Russians were making a big propaganda hit of how they were launching satellites and we were dropping rockets in a ball of fire on our launching pad. We did launch successfully, at the end of January. That was a very interesting period to live through.

—William Pickering from a transcript of an oral history in the archives of the California Institute of Technology.

In the late 1950s, there was no meeting of minds across the ideological divide.

"The country that gets a manned satellite into space first will be the undisputed master of the entire world. At the present time there is no defense against such a weapon. A satellite in a two-hour pole-to-pole orbit will pass over every part of the world every 24 hours [actually every 12 hours], and the launching of a guided missile against our cities would be a simple matter. Who is to control outer space? Russia? Or the United States?"

So wrote the editor of the *Phoenix Republic* sometime between President Eisenhower's announcement that the United States would launch satellites and *Sputnik I*'s arrival in orbit. Though more alarmist than many, the editor expressed a not uncommon fear.

Probably that fear was fueled by extracts from Soviet publications that appeared in the American press, such as the following from *Soviet Fleet*, a naval paper.

"American imperialists and their henchmen dream of using the possibility of creating an artificial satellite . . . to set up outer world bases from which it would be possible to deliver attacks against countries of the democratic camp, and to hit the selected objectives."

Amidst such rhetoric as well as the more measured and weighty criticisms of the *New York Times*, *Sputnik II* was launched on November 3, 1957. It was the second of three blows that year to America's perception of its technological supremacy. The third, a month after the second Soviet satellite, would be self-inflicted. *Sputnik II* prompted yet more questions in

Congress, more headlines, more soul-searching editorials. Congressional critics urged Eisenhower to appoint a missile czar and pour money into education. On the Monday after the launch, Senator Lyndon Johnson spent the day closeted at the Pentagon. On Thursday, Eisenhower went on national television, attempting to reassure Americans that the country was secure. He emphasized the strategic importance of the Air Force, telling his audience that the United States Air Force was as effective as missiles.[1]

But the event that presaged America's entry to the space age came on Friday, November 8, when Neil McElroy, the secretary of defense, directed the Army to prepare for a satellite launch as part of the International Geophysical Year. The Vanguard team, however, was to get the first shot.

That shot took place on December 6. The countdown went smoothly; the launch was a disaster, one that was felt all the more keenly because, unlike the Soviet launches, it took place in full view of the world. Before the entire world, the rocket lifted about two feet off the ground and then burst into flames fourteen stories high. The explosion threw the third stage and satellite clear. The satellite landed on the beach. Its bent antenna beeped to a stunned audience.

J. Paul Walsh, the deputy director for the Vanguard project, had relayed the news over the telephone to listeners at the Naval Research Laboratory. His account was succinct: "Zero, fire, first ignition—explosion."

The moment the news reached New York, there was a dash to unload Martin stock and that of other aerospace companies (though Lockheed gained). At 11:50, the governors of the stock exchange suspended trading. The next day's headlines in London included the ignominious words *Flopnik* and *Kaputnik*. Humor bolstered America, and people ordered *Sputnik* cocktails: one part vodka, two parts sour grapes.

There were to be worse failures. Astronauts and cosmonauts would die. In such a complex, unknown, and risky undertaking such disaster was

---

1. R. Cargill Hill points out that the president knew from intelligence gathered by the U2 spy planes that the Soviet Union did not have masses of ICBMs aimed at the U.S. The intelligence came from the illegal flights over Soviet territory, and Eisenhower told Secretary of State John Foster Dulles on the day before the telecast that he would not tell the nation that the United States had the ability to photograph the Soviet Union from high altitudes.

(and is) inevitable. But this one had to hurt. The space community gritted its teeth and prepared for another launch. On January 27, 1958, Vanguard came within fourteen seconds of launch. The attempt was aborted. There was a problem with the second stage. Now, though, America had only four days to wait.

Immediately after McElroy's direction of November 8, General Medaris, who headed the Army Ballistic Missile Agency (ABMA) in Huntsville, Alabama, had called Pickering, Homer Joe Stewart, and others to a meeting with himself and von Braun. The question was how to carve up the work to achieve a launch sometime toward the end of January, 1958.

Medaris assigned responsibility for the first stage of the launch vehicle to von Braun. This was the Redstone rocket, a redesigned and more powerful version of the V2. JPL was given responsibility for the satellite, the tracking stations, and the three upper stages, which would be solid-propellant rockets that the lab had developed. The entire launch vehicle was called Jupiter C.

JPL already had the tracking stations and the rockets because of its ongoing work with the Army exploring designs that would allow a missile to reenter the atmosphere without burning up. But they needed a satellite.

Pickering turned to Van Allen. They had previously talked informally about whether Van Allen's payload could be modified for an Army launch. Independently, Van Allen had talked in 1956 with staff at ABMA about an alternative, should Vanguard not be ready in time for the IGY. Van Allen would have known that delay was a possibility because Vanguard's technical director, Milton Rosen, had briefed the IGY's satellite panel about the technical difficulties with the rocket.

Therefore, once McElroy gave the army the go-ahead, Pickering sought permission from the IGY and Van Allen to prepare Van Allen's payload for an Army launch. The IGY was the easy part. It was more difficult to reach Van Allen, who was on a research vessel in the Antarctic. There Van Allen wrote in his notebook that *Sputnik* was a "brilliant achievement." His reaction (and Guier's) was in contrast to Rear Admiral Rawson Bennett's comment to NBC that *Sputnik I* was "a hunk of iron that anyone could launch."

On the deck of his cold, distant ship, Van Allen felt out of touch with the review of the U.S. program that was taking place. He was concerned that his group might miss a launch opportunity.

Van Allen was particularly worried when JPL acquired responsibility for the satellites (which became known as *Explorer*), fearing that the lab, which he perceived as very aggressive, would try to take over his experiment. His consolation was the confidence he had in Bill Pickering.

Pickering, in fact, went to considerable trouble to contact Van Allen, first with messages via the Navy. When that didn't work, Pickering recalls that someone suggested Western Union. That succeeded. Van Allen cabled his agreement that his payload should be modified for an Army launch. His assistant, George Ludwig, picked up the bits and pieces around the laboratory and, in Pickering's words, hightailed it out to Pasadena and the Jet Propulsion Laboratory.

The launch was scheduled for the end of January. This time, there was no formal prior announcement, though there were plenty of leaks. Journalists were on the alert. Shortly before the launch, Pickering's staff were telling callers that Pickering was in New York. A wire reporter, keen to be sure that Pickering was where he was said to be and not in Washington D.C. or at Cape Canaveral, turned up at Pickering's New York hotel. "Just checking," the reporter told him.

Days later, Pickering *was* in Washington—at the Pentagon with von Braun, Van Allen, senior Army personnel, and the secretary of the Army. They were waiting for the launch attempt of what would become *1958 alpha I*, better known today as *Explorer I*. High winds had delayed the shot for twenty-four-hours. But on the evening of January 31, it seemed likely that it would go ahead.

Periodically someone would call the Cape to see how the countdown was faring. At T minus 45 minutes, launch controllers halted the countdown because engineers thought there was a fuel leak. After eighteen minutes, they decided there had been a spill during fueling and wiped up the mess. The countdown resumed. The servicing structure rolled back, and its lights went out. Now a search light picked out the silver-gray missile as a Klaxon sounded in warning.

At 10:48, Jupiter C lifted off. When the Redstone finished its burn, von Braun said to Pickering, "Well, now it's your bird." The bird had apparently been injected safely into orbit. But to be sure, they had to wait. Not before the tracking station in California picked up the satellite's signal would they be confident that the spacecraft was orbiting. The only other people on the planet who could really know what that wait was like were in Kazakhstan.

Frank Goddard, at the Goddard Space Flight Center, was waiting to hear from California. Pickering kept a phone line open to him. They had predicted when they should hear the satellite. They waited as the minutes ticked past the time when they should have heard its signal. Pickering felt the glares on his back. Eight minutes after their predicted time, California confirmed that the satellite was in orbit, and the satellite completed its first orbit in the early hours of February 1.

Now von Braun, Van Allen, and Pickering were whisked through the rainy streets of Washington to a press conference at the National Academy of Sciences. They walked into a barrage of lights and microphones. "We didn't know what we were getting into," Pickering later recalled. "The place was jammed to the rafters. It was very exciting."

The news was relayed to President Eisenhower in Augusta, Georgia, where he was enjoying a golfing vacation. He said, "How wonderful."

Within minutes of the news reaching Huntsville, thousands of people took to the streets, honking car horns and carrying placards that read, "Move over *Sputnik,* our missiles never miss."

America had entered the space age.

Before *Sputnik I,* the United States had planned that its first attempt to launch a full-scale satellite would be in the spring of 1958 and that four test vehicles carrying grapefruit-sized test satellites would be launched in the autumn of 1957. There were hopes that one would attain orbit. In the event, Vanguard put its first grapefruit in orbit on March 17, the day that Frank McClure called Guier and Weiffenbach into his office to discuss how their computational and statistical approach to tracking could serve navigation satellites.

By then, the space community was growing more comfortable with the techniques of satellite tracking. Yet during 1957 they had asked themselves how they would track all the spacecraft if as many as six were to be launched each year. The question arose in 1957 as the satellite advocates tried to persuade their colleagues to endorse a continuing space program beyond 1958. Now, when TRW's *Space Log* reports that by the end of 1987 there had been 2,979 known successful satellite launches (not including those deployed from the shuttles), that concern exemplifies the adage that the past is another country.

To today's politically minded citizens, however, yesterday has familiar traces of home, namely budget battles, sniping between participants, and press relations.

By the beginning of 1956 the IGY's total budget for the satellites and tracking was $19,262,000 an amount that approximately equaled its budget for everything else. This did not include the cost of developing and building the launch vehicles. Twelve satellites had been proposed by the scientists. The administration had announced ten in July 1955. By mid 1956, the scientists could count on six but, conservatively, were selecting only four for full development within the IGY's timetable. All this took place within the context of Defense Department's budget skirmishes and the rising costs of Vanguard.

The satellite panel was warned to keep the reduction confidential lest it damage America's international prestige. Some clearly thought that this warning should not be heeded, because stories about the reduced program trickled out to the press, as did Fritz Zwicky's comments to the American Rocket Society in November 1956 that "all kinds of jealousies, bureaucracies, and buck passing" were hindering the American satellite program. Many newspapers complained that the Navy should not have got the job of launching a satellite, and others reported on delays in placing of contracts for basic components.

Relations with the press were, in general, a contentious issue between the IGY and the Department of Defense. The scientists grew increasingly irritated because, in their opinion, the Defense Department's publicity machine made the project look like a military exercise. Not a difficult job given that the Naval Research Laboratory was developing the launch vehicle and that some payloads were being prepared by scientists in defense laboratories. Nevertheless, and despite the fact that many of the university-based scientists had professional relationships at some time with the military, the scientists were determined to ensure that results of experiments were published in the open literature so that their international scientific relationships would not suffer. One can't help but wonder what Soviet scientists were going through.

Amidst these concerns perhaps the most intriguing is the one that emerges in a flurry of correspondence in early 1957 that documents that the IGY scientists feared that the Department of Defense would cancel the program once one satellite had been launched successfully.

Nevertheless, planning for four satellites continued. And the space advocates succeeded in their campaign to convince their less enthusiastic colleagues to recommend that the satellite program continue after the IGY. As late as the day before the launch of *Sputnik*, this was not certain. But *Sputnik*, of course, changed everything for the space program. Like navigation, meteorology benefited.

2

★

Meteorology

A Time of Turbulence

This dread and darkness of the mind cannot be dispelled by sunbeams, the shining shafts of day, but only by an understanding of the outward form and inner working of nature ...

First, then, the reason why the blue expanses of heaven are shaken by *thunder* ...

As for *lightning,* **it is caused when many seeds of fire have been squeezed out ...**

The *formation of clouds* **is due to the sudden coalescence ...**

—Lucretius, On the Nature of Things

Lucretius sought rational, deterministic explanations for the weather. These turned out to be wrong, but one suspects that the Roman philosopher may have guessed this for himself. He wrote that it was better to venture on an incorrect rational explanation than to submit to superstition: no sacrifices for him to propitiate the gods. And no sacrifices, except of time and effort, for those who during the past hundred years or so have wrestled to turn meteorology into a science.

For most people—farmers, sailors, or those of us going about our ordinary business—meteorology means and has always meant the weather forecast: the difference between heading for the golf course or curling up at home with a good book, between planting crops or waiting, and ultimately for some the difference between life and death. Those forecasts, dispensed in a few minutes on nightly news broadcasts, rest on the integration of a staggering amount of mathematics, physics, engineering, and computer science. In the first century B.C.E., while incorporating his own ideas with the philosophy of Epicurus and turning the whole into verse, Lucretius was at a considerable disadvantage.

Only in the nineteenth century did the modern era of weather forecasting begin. The introduction of the telegraph allowed observers to communicate to those at distance points what weather was coming their way. Such timely reporting also allowed meteorologists to plot weather maps and to develop the concept of storm fronts and cyclones. From the 1920s, radio balloons collected readings of temperature, wind speed, pressure, and moisture content, improving knowledge of conditions at altitudes in the lower atmosphere. Later, in the 1950s and 1960s, scientists took the important step of incorporating knowledge of the upper atmosphere into their understanding of meteorological conditions in the lower atmosphere, that is, they explored how the upper atmosphere affects weather at the surface.

But until the middle of the twentieth century, meteorology was only slowly breaking free of its ancient reliance on folklore and superstition. It was still more of an art than a science. Then came computers, mathematical modeling of atmospheric behavior, and weather predictions based on computer models. Gradually, it became possible to combine and manipulate observations from many different sources—from ocean buoys to Doppler radar and satellites.

Weather satellites inserted themselves into this history as best they could—not always felicitously. They were a technology in which some in the 1950s intuitively saw promise because of the unique bird's-eye view from space, but it was only in the early 1980s that the advocates of satellite meteorology succeeded in winning widespread acceptance from the meteorological community.

In the very earliest days of satellite meteorology, a few names stand out in what was a tiny, intertwined community. The first are William Kellogg and Stanley Greenfield, who in 1951 while at the RAND Corporation (consultants to the Air Force) published the first feasibility study on weather satellites. Then came Bill Stroud and Verner Suomi, who competed to have their experiments launched on one of the satellites of the International Geophysical Year. Each, after vicissitudes, flew an experiment. Stroud's failed, because the Vanguard satellite that carried it into space was precessing wildly. Stroud went on to head NASA's early meteorological work at the Goddard Space Flight Center and to argue the case for satellite meteorology at congressional hearings. Suomi's satellite produced data, and he remained in the trenches of science and engineering, making frequent forthright forays into the policy world both nationally and internationally.

There were also Harry Wexler and Sig Fritz from what was then called the Weather Bureau. Wexler, who died in the 1960s, is someone whose name in this context is often forgotten, but as chief scientist of the

Weather Bureau and an active participant in the committees planning the IGY, he was an important supporter of satellite meteorology. He was one of the scientists arguing persuasively in the face of Merle Tuve's doubts that the IGY should include a satellite program. And Wexler was a staunch ally of a belated attempt by Verner Suomi to participate in the IGY, drumming up support for Suomi from eminent meteorologists like Kellogg at RAND.

Fritz worked for Wexler. When the Weather Bureau set up a satellite service, Fritz was its first employee. He was assigned office space in a cleaned out broom cupboard. There, undaunted by the Vanguard failures and the modesty of his office space, Fritz worked with NASA on the first American weather satellite—TIROS. Both Wexler and Fritz were consultants for Verner Suomi's IGY experiment.

Fritz recruited Dave Johnson,[1] who, like Suomi, became an outspoken proponent of satellite meteorology. Johnson eventually headed the satellite division of what, after several bureaucratic incarnations, was to become the National Oceanic and Atmospheric Administration.

Except for Kellogg and Greenfield, these men worked in the civilian world but also made forays into the "black" world of defense projects, namely the Air Force's Defense Meteorological Satellite Program. The Air Force was an important player in the history of satellite meteorology, developing both engineering and analytical methods for interpreting satellite imagery. And the participation of people like Johnson in both worlds provided a conduit, albeit of limited capacity, for technology transfer from military to civilian satellites. The story of this important part of the history of satellite meteorology—the way that the defense and civilian worlds intermixed—will have to wait until all the relevant documents are declassified.

---

1. At least two other people should be mentioned in connection with the early days of civilian weather satellites. They are Robert White and Fred Singer. White, who retired in 1995 as president of the National Academy of Engineering, was the head of the Weather Bureau and administrator of NOAA in the 1960s and 1970s. He was an influential supporter of satellite meteorology. Fred Singer, who was a member of the Upper Atmosphere Research Panel and worked for a while at the Applied Physics Laboratory, oversaw important engineering advances to TIROS. He was, says White, and is, a fascinating person and a maverick. His most recent provocation to the scientific community is a disbelief in the human contributions to global warming. In the early 1960s, several other people joined the new field of satellite meteorology, and anyone interested in learning about the technology in detail should see Margaret Eileen Courain's Ph.D. thesis, *Technology Reconciliation in the Remote Sensing Era of US Civilian Weather Forecasting,* Rutgers University (1991).

Despite the limitations imposed by not having a full understanding of the interplay between civilian and defense projects, some broad aspects of the history of satellite meteorology are clear. It is a more complicated story than that of satellite navigation, mainly because it is the story of a technology being developed for a field that was still transforming itself from art to science.

One of the most outspoken and energetic participants in the field's history was Verner Suomi, of the University of Wisconsin in Madison. Some have called him the father of satellite meteorology.

In 1992, Dave Johnson, then working for the National Research Council of the National Academy of Sciences, recalled a meeting of the world's leading meteorologists in 1967 when they were planning an international effort, known as the Global Atmosphere Research Program, to study the atmosphere. GARP eventually got underway in the late 1970s. Suomi's task was to summarize the specifications that weather satellites would have to meet in order to fulfill GARP's research goals. Johnson said: "We threatened to lock Vern in a room and not to let him have food or drink until he'd written everything down. We didn't, of course, but he hated writing, and we had to keep an eye on him."

Suomi's colleagues were wise to put pressure on him. During late 1963 and early 1964, when Suomi spent a year in Washington D.C. as chief scientist of the Weather Bureau, he claims to have written only four memos—which may be the all-time minimalist record for a bureaucrat.

One of GARP's roles was to set research priorities given what were then the comparatively new technologies of high-speed computing, mathematical modeling, and satellites. Those priorities give a sense of the immensity of the task facing meteorologists.

The priorities were:

- Atmospheric composition and structure;
- Solar and other external influences on the earth's atmosphere;
- Interaction between the upper and lower atmosphere;
- Interaction between the earth's surface and the atmosphere;
- General circulation and budgets of energy, momentum, and water vapor;
- Cloud and precipitation physics;
- Atmospheric pollution;
- Weather prediction;

- Modification of weather and climate (no longer popular);
- Research in sensors and measuring techniques.

A study of these topics would need the "observation heaped on observation" that Sir Oliver Lodge spoke of in his lecture about Johannes Kepler: some observations were to be made by radar, others by airline pilots, weather balloons, and ground-based instruments. And some, of course, would be recorded by satellites.

Despite the vibrancy of meteorological research typified by plans for GARP, it was clear by 1967 that persuading the wider meteorological community—both line forecasters and many research meteorologists—to accept data from satellites would be an arduous task.

Many of the important steps to acceptance were choreographed, in part at least, by Suomi or Johnson and the groups that they headed. Neither man was shy in his advocacy of the technology. Johnson, in fact, threatened on one occasion to "blow his stack" with his boss, whom Johnson felt was hostile to satellite data. None of the advocates of satellites could afford too many niceties. The money spent on weather satellites prompted resentment from many. And there were reservations and criticisms about satellite meteorology.

Part of the opposition lay, as always, in suspicion of a new technology. But part of it was due to the technology's acknowledged limitations, which were (and are) imposed by the nature of satellite observations. Satellites do not directly measure the meteorological parameters—temperatures, pressures, wind speeds and moisture contents at as many latitudes, longitudes, and altitudes as possible—that are essential for computer models and any quantitative predictive understanding of the atmosphere's behavior. Instead, satellites "see" visible and infrared radiation welling up from the earth. Meteorologists thus have either images or radiometric measurements as their raw data, and from these they must infer quantitative meteorological parameters. The inferences are not easy to draw. They call for considerable knowledge of atmospheric physics and chemistry and rely on clever mathematical manipulations of the equations describing atmospheric behavior.

Images rather that radiometric measurements came first in the history of meteorology satellites. Kellogg's and Greenfield's study of satellites for "weather reconnaissance," which was carried out before numerical weather prediction had become central to the future of weather forecasting, envisaged that spacecraft would carry still cameras aloft. These would photograph cloud cover, and meteorologists would then study the cloud types and distribution in a qualitative attempt to gain insight into atmospheric behavior and thus improve weather forecasting. In the course of their study, Kellogg and Greenfield posed some of the important questions that would preoccupy early satellite meteorologists. These were:

- How could you tell which bit of the earth the camera was looking at and thus where the cloud cover was?
- How could you tell what type of clouds you were looking at and what their altitudes and thicknesses were, and thus what significance they had to a developing weather system?
- How could you get the information to line forecasters in a timely fashion? It would not be much use telling a ship that there had been an eighty percent chance of a storm yesterday. The launch of the first weather satellite—*TIROS I* (for thermal infrared and observing system)—in April 1960 confirmed that these were all tough and legitimate concerns.

Nevertheless, *TIROS* showed for the first time what global weather patterns looked like. The promise inherent in the technology was there for all to see in grainy black and white. But it convinced only those who already believed. Succeeding satellites in the TIROS and improved TIROS series carried gradually more sophisticated instruments, each of which slowly took satellite meteorology closer to wide acceptance.

One such class of instruments—known as sounders—were first developed by Johnson's group in the 1960s. Sounders measure temperature and, more recently, the moisture content of the atmosphere at different altitudes and in places where direct measurements with, say, a thermometer are not possible—over oceans, for example, where much of the weather develops. They are important for near-term predictions of severe weather such as thunderstorms.

The sounder relies on inferences made from radiometric readings at different frequency ranges in the infrared portion of the spectrum and on its operators' detailed knowledge of atmospheric chemistry and physics. Inevitably, there is greater inaccuracy in the values of temperature and moisture content taken from satellite sounders than from direct measurements of the same parameters. And so modelers have, for the most part, not liked to rely on data from satellite sounders. A notable exception is the European Center for Medium Range Weather Forecasting, which has taken the lead in finding ways to extract from satellites the information that is needed for computer models. By the early 1990s, the center was saying that satellite soundings had extended useful predictions from five and a half to seven days in the Northern Hemisphere and from three and a half to five days in the Southern Hemisphere.

While Johnson's group developed the first sounder, Suomi came up with the idea for the spin-scan camera, which flew for the first time in 1966. Although this class of camera was to become a crucial meteorological instrument, Suomi was told by a colleague ten years after it first flew that if submitted as part of a Ph.D. thesis, it would not merit a doctorate.

Thus satellites were not entirely welcome participants in meteorology. Far more welcome were the new high-speed computers and John Von Neumann's conviction that with sufficient computational power one could model the atmosphere's behavior and predict the weather.

The idea for such numerical weather prediction was proposed first in 1922 by Lewis Richardson. He tested his idea by feeding meteorological data that had been collected at the beginning of International Balloon Day in May 1910 into mathematical models describing atmospheric behavior. He compared his numerical predictions with the data collected during the day and found no agreement. Discouraged, Richardson concluded that to predict the weather numerically one would need 64,000 mathematicians who would not be able to predict weather conditions for more than seconds ahead; they would, in effect, be "calculating the weather as it happened."

In the thirty years following Richardson's depressing experience, much changed, including improved understanding of the physics of the atmosphere and mathematical analysis of its behavior. Thus, when the technology of computing emerged, modelers set to work, weaving the basic physical laws into models mimicking the behavior of the atmosphere. And the computers took over the calculations. Initially, the models represented only surface events in small regions. Subsequently, modelers incorporated the influence of the upper atmosphere on weather at the surface.

There are now many models—global, hemispheric, regional. Some are mathematical behemoths constructed from thousands of equations. Some give short-term weather predictions, while others look up to two weeks ahead—so-called medium-term forecasts. Yet others make forecasts, extremely controversial ones, far into the future as climatologists explore climatic change.

All, however, devour numbers—values of temperature, pressure, etc. And because the early satellites did not supply the quantitative data that the models required, there was tension between computer modelers and satellite advocates. Both groups, after all, were seeking scarce public funds for expensive technologies.

In 1969, ten years after the first meteorological payload was launched, the National Academy of Sciences wrote, "... numerical weather prediction techniques demand quantitative inputs, and until weather satellites are able to generate these, their use in modern meteorology will be at best supplementary."

Nearly thirty years later, the technologies have become more compatible and weather satellites have obtained a secure place in meteorology. The Air Force, NOAA, NASA, and academic groups like that of Suomi's at Wisconsin have done what they can to extract meteorological values from unprepossessing streams of satellite data and, importantly, to make this information compatible with observations from weather balloons, radar and surface instruments. Yet, says Johnson, considerably more information could be extracted from the meteorological satellite data.

Weather satellites gather their data—images and soundings—from two different types of orbit: polar and geostationary. Like *Transit,* a weather satellite in polar orbit follows a path that takes it over the poles on each orbit, while the earth turns through a certain number of degrees of longitude in the time it takes the satellite to complete one orbit. Thus polar-orbiting satellites, if they have a wide enough field of view to either side of the subsatellite point, provide global coverage. Their altitude, and thus how long they take to complete an orbit, is chosen so that the satellite will "see" all parts of the earth once every twelve hours.

To be truly useful, however, weather satellites need to occupy a special kind of polar orbit, known as sun-synchronous. Sun-synchronous orbits are chosen so that the satellite maintains the same angular relationship to the sun, which means that the satellite will be above the same subsatellite point at a given time of day. Its readings are then consistent from day to day. The timing of the orbit is chosen so that the satellite readings are available for the computer prediction models, which are run twice a day.

If the orbit is to maintain the same angular relationship to the sun throughout the year, it cannot remain fixed in space. But orbits are not, of course, fixed. They respond to the earth's gravitational anomalies. Mission planners achieve sun-synchronous orbits by exploiting the known effects of the earth's gravitational field. They select inclinations and altitudes that result in the orbit moving in such a way that the satellite's sun-synchronous position is maintained. The consequences of the natural world that the Transit team had to understand and to compensate for can thus be exploited usefully by those planning the orbits of weather satellites.

The laws of physics result, too, in the existence of the extremely useful geostationary orbit. A satellite at an altitude of about 36,000 kilometers takes twenty-four-hours to complete an orbit. If the orbit has an inclination of zero degrees, that is, the plane of its orbit is coincidental (more or less) with the plane of the equator, then the satellite remains above the same spot on the earth. Thus, the satellite is with respect to the earth for all practical purposes stationary and can view the same third of the earth's surface while the weather moves underneath it. Suomi's spin-scan cameras were designed for this orbit. Geostationary orbits were also to prove of critical importance to communication satellites, and Suomi's spin-scan camera was first launched aboard a satellite designed by one of the fathers of communication satellites—Harold Rosen.

While they are crucial to the beginnings of satellite meteorology, the issues mentioned so far scarcely scratch the surface of the history of weather satellites. There was also an important battle in the early 1960s between NASA and NOAA's forerunner about the technology of the satellites to replace TIROS and about who would pay for operational satellites. Finally, an improved version of TIROS was selected, and NASA developed the alternative proposal, a more experimental satellite series dubbed Nimbus.

White recalls, "On the same day I was sworn in as chief of the Weather Bureau, Herbert Holloran, the assistant secretary for science and technology, took me to one side and said we have to make a decision about Nimbus. The issue was would we be willing to use Nimbus as our operational satellite. The cost would have been two to three times the cost of using TIROS. This was important to the weather satellite program. If we had followed Nimbus, the cost would have skyrocketed, and maybe we wouldn't have got the money from Congress. We decided on the basis of cost to go with TIROS. I think that was the right step."

Even from a technical standpoint, the history is not straightforward. There was no single event, such as Guier and Weiffenbach's tuning into Sputnik's signal, from which the story unfolds. Nor was there one clearly defined technical goal such as that of the Transit program—locate position with a CEP of one tenth of a mile. All of the physics and engineering that went into the Transit program were harnessed to meet that goal and were refined to enable the subsequent improvements in the system. In the field of meteorology, satellites were just one tool wielded to learn more about the atmosphere, and no one really knew what needed to be learned as is apparent from the breadth of The Global Atmosphere Research Program's aims. It is, therefore, not surprising that meteorology satellites took longer than navigation spacecraft to find acceptance.

In further contrast to Transit, there was no single group, like the Navy's Special Projects Office, that wanted weather satellite technology. Even the Weather Bureau, outside of Johnson's group, was unenthusiastic. Further, no single group, like the Applied Physics Laboratory's Transit team, was central to the development of weather satellites. True, the Air Force, backed by sundry laboratories and consultants such as the RAND Corporation, was interested from early days, but once the IGY's satellite program was announced, more scientists became involved, including Verner Suomi and Bill Stroud. After the launch of *Sputnik I,* the Advanced Research Projects Agency sponsored the TIROS program, which NASA took over when that agency opened its doors in October 1958. Industry, including companies like RCA, took a hand, and, of course so did the Weather Bureau.

If the professionals were slow to accept meteorology satellites, the lay audience was intrigued by the potential of a spacecraft's global view, and popular articles appeared in the newspapers of the 1950s speculating on the importance of satellites for weather forecasting. They pointed out that only satellites would be able to provide comprehensive and frequent readings over the approximately seventy-five percent of the earth's surface that is covered by ocean.

Since the first TIROS went into orbit, the United States has launched more than one hundred meteorology satellites. Now the countries of the former Soviet Union, Europe, Japan, the People's Republic of China, and India maintain meteorology satellites. All contribute to the global economy by improving forecasts for agriculture and transport, and to safety by monitoring severe weather such as hurricanes and allowing more timely and accurate predictions of where they will make landfall. It is unlikely, in the U.S. at least, that a hurricane will ever kill more than 6000 people, as did the hurricane that struck Galveston, Texas, in 1906. It has taken more than three decades, but weather satellites are now living up to the popular expectations of the 1950s.

The Bird's-Eye View

It is obvious that in observing the weather through the "eye" of a high-altitude robot almost all the quantitative measurements usually associated with meteorology must fall by the wayside.

—From a Project RAND report: *Inquiry into the feasibility of weather reconnaissance from a satellite* (1951), by William Kellogg and Stanley Greenfield.

During World War II, Japanese paper balloons floating on currents in the upper reaches of the lower atmosphere carried incendiary bombs across the Pacific to the United States. They caused some forest fires, which were quickly extinguished. Censorship kept news of the few fires from the public, and thus the balloons did not precipitate the widespread consternation that Japanese strategists had hoped for.

William Kellogg and Stanley Greenfield were intrigued by the story of these balloons and were impressed by the knowledge of the atmosphere that such a campaign had needed. The two men worked for a newly formed group of technical consultants known as the RAND Corporation.[1] The interest the two men had in high-altitude balloons, which they believed might make good platforms for photo reconnaissance and intelligence gathering, evolved eventually into a conviction that satellites would provide a good platform from which to observe the weather. Their early conceptual work on "weather reconnaissance" became part of TIROS, and in 1960, the American Meteorological Society presented Kellogg and Greenfield with a special award.

The RAND Corporation was an ideal place for Kellogg's and Greenfield's work. The organization was formed in 1948 when the U.S. Air Force, previously the Army Air Forces, became a separate branch of the armed services. It grew from Project RAND, which the Douglas Corporation set up immediately after the Second World War to evaluate advanced technology for the Army Air Forces' missions. RAND made its

---

1. See *RAND's Role in the evolution of balloon and satellite observation systems and related US space technology,* by Merton E. Davies and William R. Harris, published by the RAND Corporation.

first analysis of the technical feasibility of satellites for the Army Air Forces in 1946.

In 1948, the government assigned responsibility for satellites to the Air Force, which appointed RAND to manage the work. RAND subcontracted studies on guidance, stabilization, electronic reliability, communications, space reconnaissance systems, and space power systems to companies such as Westinghouse, Bendix, and Allis Chalmers. In 1951, RAND published classified studies incorporating industry's and its own work. These studies analyzed potential missions as well as engineering design, the political implications of the technology, and the potential of satellites for science. Later, when the U.S. space program got seriously underway following the launch of the first two *Sputniks,* the content of these reports would be incorporated into the early classified satellite programs, such as the Discoverer series (on which APL had a transmitter).

RAND's main conclusion in 1951, however, was that satellites had potential as observation platforms for reconnaissance. In the same year, Kellogg and Greenfield completed a study on the feasibility of satellites— referred to as satellite missiles or satellite vehicles—for "weather reconnaissance."

Weather and reconnaissance satellites proved to be a hard sell, but of the two it was reconnaissance satellites that first won the administration's backing. In winning that backing, reconnaissance satellites precipitated, by the tortuous paths that characterized the Cold War, the U.S. space program, including—eventually—meteorology satellites.

Initial opposition to reconnaissance satellites focused on their technical limitations. In those days photography from high altitudes offered spatial resolutions of the order of hundreds of feet, knowledge obtained from high-altitude rocket photographs (thirty, forty-five, and sixty-five miles), which the Navy had been the first to take. This resolution was much poorer than what could be obtained from cameras on aircraft, and those who had developed expertise in photo interpretation during World War II were scornful of the technology's capabilities.

But prompted by the need for better intelligence, both to prevent surprise attacks and to monitor arms-control agreements, the Eisenhower administration proposed low-level funding for a reconnaissance satellite program in fiscal year 1956 (that is, for funds available from October 1955). The money authorized was as follows: fiscal year 1956—$4.7 million; 1957—$13.9 million; 1958—$65.8 million.

Air Force historian R. Cargill Hall says that the critical factors in winning backing for the development of a reconnaissance satellite were that satellites offered a photographic platform that could not be seen by the naked eye or detected by radar sensors, and if satellites were detected, they would be too far away to be shot down.

These were not advantages that the Soviets were likely to appreciate. When President Eisenhower suggested in 1955 that there be an "open skies" policy, allowing overflight of Soviet and U.S. territory in order to verify disarmament agreements, Khrushchev had dismissed the idea, calling it "licensed espionage." Khrushchev viewed the policy as an attempt to gather information on potential military targets, and that, indeed, was one of its purposes. So the prospect of American spy satellites could be guaranteed to provoke Soviet animosity. For these and other reasons, argued the historian Walter McDougall, the Eisenhower administration wanted a civilian satellite launched first to establish the precedent of the freedom of space. In an article in *Prologue, Quarterly of the National Archives* (spring 1996), Cargill Hall confirms this view, arguing that the IGY enterprise effectively was made into a stalking horse for military reconnaissance satellites.

It was not a cheap stalking horse—nearly $20 million initially for the satellites alone—but it was effective. The program played host to enough political tensions, technical difficulties, protests from the Army, and criticisms of the Eisenhower administration's Vanguard decision to keep the eyes of the world focussed on the IGY. Thus it was that IGY became the forum in which the first satellites were launched.

To the disappointment of Kellogg and Greenfield, weather satellites languished during the discussions about reconnaissance. Even though the resolution was poor, satellite images, they believed, offered something of potentially great importance to meteorologists—a "synoptic" picture, assembled from several individual photographs, showing cloud patterns over a large expanse of the earth at one time.

The proposal Kellogg and Greenfield made in 1951 had to wait seven years until the Advanced Research Projects Agency was prompted by Sputnik into finding applications for satellites. William Kellogg was then appointed to head an ARPA panel drawing up specifications for a meteorology satellite.

Kellogg turned to his and Greenfield's early ideas, which, as in the case of reconnaissance satellites, owed much to rockets and high-altitude balloons. Images from rockets enabled meteorologists to begin developing analytical techniques that made sense of the bird's-eye view of the Earth, while balloons served as test beds for meteorological instruments and cameras. The balloons could carry more weight aloft than could early satellites.

In the late 1940s, high-altitude balloons were already a versatile technology for a variety of applications. William Kellogg had worked on a project for the U.S. Atomic Energy Commission investigating their potential for monitoring the dispersion of radioactive particles from atomic tests; the Japanese, as we saw, had used them to carry incendiary bombs; and the Photo Reconnaissance Laboratory at the Wright Field in Ohio established that balloons made a stable platform for aerial photography.

At about the same time, in January 1949, the *Bulletin of the American Meteorological Society* published an article by Major D.L. Crowson, "Cloud Observations from Rockets." Crowson suggested that even low-resolution imagery from high altitudes would improve weather forecasting.

Given this background, it is not surprising that Kellogg and Greenfield decided to pursue the idea of using high-altitude balloons first for photo reconnaissance and then for meteorological research. This work gave them the information they needed about optical systems for their 1951 report on weather reconnaissance from a satellite. What they wrote was by no means a detailed engineering proposal, but it tackled conceptually for the first time the elements of a meteorological satellite carrying a camera operating in the visible portion of the electromagnetic spectrum.

They asked, Can enough be seen from an altitude of 350 miles to enable intelligent, usable weather observations to be made, and what can be determined from these observations?

From analyses of photographs taken by rockets, they decided that a resolution of five hundred feet was necessary if all the useful cloud structures were to be identified.

They assumed that the camera would mechanically scan to build a photograph of a wide enough area—a 350-mile swath around the sub-satellite point—to be of use. Once they had specified the minimum resolution, they discussed the aperture, illumination, exposure time, and focal-length-to-aperture ratio needed to achieve a given contrast. The satellite, of course, would not be taking carefully posed and cunningly lit photographs but would have to operate in whatever conditions nature decreed. So the camera and optical system had to be chosen to provide usable photographs in a variety of conditions.

Photographs, they recognized, would not yield the quantitative information that meteorologists needed. It would be impossible to do more than make intelligent guesses at temperatures and pressures. But in those days, before numerical weather prediction, these limitations did not seem as great an obstacle as they would shortly become. Thus, unknowingly, Kellogg and Greenfield put their finger on what was to be the main problem in winning acceptance for satellite meteorology. Analysts, they wrote, would have to make the most of the visible aspects of the weather in building their weather charts; clouds, being the most visible aspect of the weather, would be important.

Rocket photographs had already given an inkling of the inherent problems. In pictures taken by cameras on Thor and Atlas, it was difficult to tell whether areas of uniform greyness were cirrus clouds (wispy, high-altitude clouds formed of ice particles), tropospheric haze, or an artifact of the optical system resulting from the wavelengths accepted. All was not bad news, however, because more cloud patterns had been apparent in rocket photographs than had been expected.

In an attempt to get a feel for how accurately one might forecast weather from satellite photographs, Kellogg and Greenfield had estimated the synoptic situation from photographs taken during three rocket flights and had then made a forecast and compared it with records of actual weather on the day in question. Encouraged by the results, the two concluded that "combined with both theoretical knowledge and that gained through experience, an accurate cloud analysis can produce surprisingly good results."

By the end of 1959, anticipating the launch of *TIROS* in 1960, leading researchers met in Washington to discuss cloud research, a field known as nephology. Harry Wexler, then chief scientist of the Weather Bureau, pointed out that until the late nineteenth century, clouds had been almost the only source of information about conditions in the upper atmosphere, but that with the advent of balloon soundings, interest in nephology had declined. (Now clouds are recognized to be of crucial importance in meteorology, particularly in climate studies, and meteorologists are asking such basic questions as "What is a cloud?")

Sig Fritz, who worked for Wexler, spoke of the strong sense researchers had that they would not know how to interpret cloud photographs. This was a problem that Kellogg and Greenfield had foreseen in their 1951 report, and they had recommended that in preparation for satellite images, meteorologists build a comprehensive atlas of clouds as seen from above.

TIROS would soon begin that process.

The conception and birth of the TIROS satellites were difficult. First, in late 1957, the secretary of defense, Neil McElroy, agreed that a new agency—the Advanced Research Projects Agency (ARPA)—would have responsibility for key defense research and development projects. ARPA officially opened its doors on February 7, 1958, and weather satellites became one of its projects.

Immediately, Kellogg was asked to define the specifications for a weather satellite, which he did with the help of people like Dave Johnson. In the meantime, RCA had submitted a proposal to the Army Ballistic Missile Agency for a reconnaissance satellite as part of the Redstone missile program.[2] The Department of Defense decided that the images from this satellite would not be good enough for intelligence gathering, and it therefore became TIROS, modified to incorporate the conclusions from Kellogg's group at ARPA. ARPA managed the TIROS project until the National Aeronautics and Space Administration took it over in October 1958.

The newly formed NASA was a powerful organization, and it could expropriate groups and organizations. One of the groups that the agency wanted was that headed by Bill Stroud, of the Army's Signal Corps of Engineers, which was working on a camera for the IGY's cloud-cover experiment. After some altercations with military bureaucrats, Stroud was able to transfer to NASA in 1959, and he headed the agency's meteorology branch at the Goddard Space Flight Center.

So TIROS had its roots in a spy satellite proposed to the Army by RCA but became a weather satellite managed first by ARPA and then by Stroud's group at NASA.

The satellite's optics had a field of view four hundred miles on a side. It carried a small Vidicon TV tube, selected because of its light weight.

2. *Technology Reconciliation in the Remote Sensing ERA of US Civilian Weather Forecasts,* Courain, Rutgers University.

The images recorded were poor because the camera's electron-beam scan was not well controlled. Also, recalled Verner Suomi, who soon became involved with the TIROS program, "We didn't know what the devil the damned thing was looking at. There were some problems as to what time the pictures were taken, and the spacecraft was spinning like a top. Where the devil was north? That caused major problems."

Sean Twomi, of the Weather Bureau, soon identified the cause of one of the problems. The spacecraft needed to spin to remain stable in orbit, but the spin axis was tilting because of interactions between the spacecraft's electrical systems and the earth's magnetic field. Thus it was difficult to know where the cameras were pointing. Once Twomi had identified the problem, the Air Force developed magnetic stabilization to control the orientation of the spin axis, though it was some time before the engineering problems of orientation and stabilization of weather satellites were fully solved.

Despite such difficulties, those involved, like the engineers and scientists at the Applied Physics Laboratory, had an overwhelming sense of being pioneers. Bob Ohckers, an engineer who worked on the TIROS program at RCA and later moved to Suomi's lab at the University of Wisconsin, recalled, "In those days, there were no cutesy requirements, no quality control or oversight. Everything was experimental. If we had a failure, we would try to keep it from the contractor, particularly from Thomas Haig, who headed the Air Force's weather satellite program (Haig also joined the University of Wisconsin, where he and Suomi spent some time feuding). Haig would try to ferret out what was going on. We were told to tell him nothing. The whole group was working with rolled up sleeves and screwdrivers."

Despite the limitations of the TIROS satellites, both in terms of the data they collected and of the analytical techniques available for data processing and interpretation, the first images returned to Earth were tantalizing.

In 1964 Suomi gave a lecture to children at a local school. A copy of the speech, recorded by some attentive listener, was in an old filing cabinet in the basement of the building where Suomi worked at the University of Wisconsin. He told them how unsophisticated and crude the satellites were. But he also told them that during one orbit of TIROS they had identified two hurricanes in the southern hemisphere before they had been spotted by ships or weather stations. He showed them bright areas of cloud, telling them that this meant that the clouds were thick and high and represented an enormous thunderstorm, but that they only learned these things after the event. "We have much to learn about how to apply these pictures. The future depends not on the hardware, not on the gadgets, not on the software, but on individuals applying their knowledge to this very challenging problem [of interpretation]."

With the formation of NASA, the defense and civilian meteorology satellite programs went their separate ways. Until 1965, the Defense Meteorology Satellite Program (DMSP) was one of a suite of satellites controlled ultimately by the CIA—an indication of how intertwined the missions of reconnaissance and meteorology satellites were. Then, in 1965, control of funding for DMSP was transferred to the Air Force. DMSP remained secret until John McLucas, the under secretary for the Air Force, made the program public in 1973.

Since 1958, when the two programs diverged, the civilian and defense meteorological worlds have resisted all political efforts to reunite them. Meetings about a merger between the two were, says one participant, like arms control negotiations of old, where people developed their arguments as to why arms control was not possible.

Suomi, who like many of his colleagues worked on both sides of the fence, recalled that he would see the same work being done twice. "Part of my activities here [at the University of Wisconsin], which were classified then, was to put the heat budget experiment on a military satellite. The thing that was interesting was that many of the things that the civilian program utilized were actually developed for the military. What interested me as a party to both was that I saw one part of the RCA building which was classified and I saw another part of the building not classified. Both parts were classed as development, but really they were only one development. Someone made a lot of money on that."

Johnson, too, saw duplication, but said that interactions between the two worlds could work well when individuals in the military program were cooperative. On one occasion Johnson wanted to fly a new tube, and the DMSP allowed one of its own tubes to be replaced by Johnson's. Many of the DMSP people would also, says Johnson, do what they could to move expertise from the "black" world. But technology was not always transferred. Suomi recalled that DMSP effectively equipped spacecraft with "exposure meters" so that they could photograph clouds by moon-light in the visible part of the spectrum. Asked if such technology would have helped civilian weather satellites and why it was not transferred, he answered, "Yes" and "I don't know."

Though declassification may clarify some of these questions, it is unlikely to revise substantially the pioneering role that Verner Suomi played.

Keep it Simple, Suomi

14

> We just got along like brothers. His interests and personality were what was missing from mine. I tend to dream a lot and jump to conclusions. He was thorough and patient. How hard we worked.
>
> —Verner Suomi, talking about his friend and former colleague, Robert Parent

> As I look back, I can see all the trial and error. I don't want to use the word comedy.
>
> —Leo Skille, an electronics technician who worked with Suomi and Parent from 1959

It rained on the morning of Verner Suomi's memorial service. Deep-throated Midwestern thunder played its inimitable summer accompaniment. And as one might expect among people gathered to celebrate Suomi's life, there were questions—asked humorously, sadly—about just who it was that was responsible for the weather.

Ponytailed graduate students joined family, visiting scientists, and colleagues from Suomi's own academic home at the University of Wisconsin, Madison, in the Lutheran church on the edge of campus. Soon organ and trumpet released the opening notes of Giuseppe Torelli's Sonata in D Major: lively notes that grew thoughtful, a little sad, and again lively.

Suomi's son Stephen spoke of a man who could be stubborn, even too stubborn, but who would fake footprints and sled marks in the snow to convince him for one more year that Santa Claus existed. The minister spoke of a great scientist, but here he missed Suomi's essential strengths, which lay in his engineering ingenuity, in his experimenter's determination to collect data, and in the stubbornness that his son recognized.

Suomi's engineering ability, his "practical eye," helped him to see how to build instruments that took advantage of the spatial and geometric relationships between orbits, the rotating Earth, and the satellites, which were themselves spinning. Time and again he watched as instruments embodying his engineering concepts were launched. During interviews in the late spring of 1992, Suomi recalled that "what was exciting was to see this big rocket with 'United States' written on it, and your gadget was in the nose cone."

It was Suomi's stubbornness that placed the instruments in those rockets and kept his interest in satellite meteorology alive during the two decades it took him and other pioneers to find ways of extracting useful information from satellite data and to convince the wider meteorological community of the information's validity. Verner Suomi held his own in the competitive and dynamic arena of meteorology. He knew quite well that he was stubborn, and said of himself, "I just yell and holler until I get my way."

Building things was important to Suomi throughout his life. He was fascinated as a teenager by radio and flying. To his generation—the teenagers of the late 1920s—these comparatively new technologies were what computing is to the scientifically inclined youngsters of the late 1990s. It was almost obligatory to build your own radio set, which Suomi did during his summer holidays. He took a car mechanics course at school and worked on an aircraft engine as he trained to be a teacher at Winona State College in Minnesota.

As Suomi was contemplating these memories in his office, he recalled that some years past he had met an old girlfriend, someone from the days before he met his wife, Paula. This woman had commented on his patience with things but lack of patience with people.

"I thought a lot about that; it had a big affect on me and was a milestone. Since then, I've tried to be patient with people. Whether I was successful or not, well that's another story. The thing is, it's easy to work with things, they don't talk back."

Given this aspect of Suomi's character, it makes sense that at the age of twenty-seven he gave up teaching high school to take a degree in meteorology at the University of Chicago.

Chicago's meteorology department was run at that time by one of the giants in the history of the subject, Carl Gustav Rossby. Reportedly, it was after learning from Rossby of Lewis Richardson's failed attempt to make a numerical weather prediction that John Von Neumann concluded that weather forecasting was an ideal problem for the new "high-speed" computers.

Suomi had learned a little about meteorology and weather forecasting while he was learning to fly but had dropped the subject after he had received his license. (As he talked about flying he wistfully remarked that he would like to fly "one of those big planes.") At Chicago, he found Rossby to be an original and fascinating person, someone who "could talk you out of your shirt and make you think you had a bargain."

Suomi also met Reid Bryson, who established a meteorology department at the University of Wisconsin, one that became among the best in the country. Bryson asked Suomi to join him in developing the department, which subsequently proved to be a source of discord between the two men. Suomi and Bryson developed a complicated relationship that incorporated friendship and professional rivalry. They had both moved to the University of Wisconsin in 1948, where Suomi completed the doctorate he had begun in Chicago, earning his degree at the age of thirty-eight in 1953.

In many ways, Bryson's and Suomi's rivalry typified the greater rivalry that existed in meteorology between atmospheric modelers and advocates of satellites. Suomi, of course, was the satellite advocate, interested in improved weather forecasts; Bryson was more interested in computer modeling and climate. With these interests, both placed themselves out of what was then the mainstream of meteorology—Suomi because satellite instrumentation was not providing what meteorologists needed, and Bryson because in the late 1950s and early 1960s climate modeling and prediction were not quite respectable. In the pursuit of more accurate weather forecasts, both for long-range and short-term predictions, tension over resources and staffing arose between the two men. Neither was content to rely only on traditional instruments and weather maps. Both satellite technology and bigger and better computers on which to run more detailed models cost money and call for expensive expertise, and local tensions mirrored these at a national level. The fierceness of the disagreement at all levels was fueled by genuine scientific disagreement about the correct way to move meteorology forward.

As a general rule, the modelers were theoreticians at heart, while the satellite advocates had the souls of experimentalists. Of course, experimentalists had to understand theory and theorists had to incorporate the results of experiment, but there seemed and seems to be a difference between the two groups that can set the two sects at one another's throats.

The difference was manifest between Bryson and Suomi. Bryson worked on theories of climate. Suomi wanted to design instruments to learn more about what were then the largely unexplored regions of the atmosphere in the Southern Hemisphere and at higher altitudes, regions which, in an interview with William Broad of the *New York Times,* Suomi called the ignorosphere.

In the end, of course, meteorology proved to be big enough for both of them and to need both of their approaches. And they continued to live as neighbors in the houses they had helped one another build.

During the 1950s, Suomi formed another important professional relationship, one that was synergistic. This was with Robert Parent, from the university's Department of Electrical Engineering. Their first collaboration was in the design and construction of an experiment known irreverently as Suomi's balls, and more properly by the title of their proposal to the International Geophysical Year, which was "satellite instrumentation for the measurement of the thermal radiation budget of the earth."

This property of the earth—its thermal radiation budget—is a crucial aspect of meteorology. It may even be the most basic factor influencing the world's climate. Thus, the radiation budget is of more interest to basic research than in the applied field of weather forecasting. And there is some irony in the fact that Suomi, who became known for designing satellite instruments for improving short-term weather forecasting and who had the differences of opinion that he did with Bryson, should have been the first to put forward a proposal to measure the earth's radiation budget.

The radiation budget is the driving force behind both atmospheric and oceanic circulation. Very simply stated, it is the balance between the incoming solar radiation and the outgoing radiation from the earth. Over time and over the whole earth, the incoming and outgoing radiation streams balance one another. The importance of the radiation budget, however, lies not in this balance but in the way that the balance of the incoming and outgoing radiation varies with geographical location and time.

The outgoing radiation falls into two broad categories: that which is reflected from the atmosphere and the surface without any change in wavelength, and that which is absorbed and reradiated at longer wavelengths from the surface and the atmosphere. The amount of radiation reflected by some structures, such as ocean, cloud, or icecap, is known as the albedo of that surface. Meteorologists know now that the mean albedo of the earth is thirty percent, but that individual surfaces can have albedos ranging from ninety percent for dry snow at high latitudes to fourteen percent for dark, moist soil. Since the albedo is greatest at the poles, and concomitantly, absorption and reradiation are lowest at these latitudes, heat flows to the poles from the equator, driving the atmospheric and oceanic circulation without which there would be no weather.

Many types of atmospheric behavior and processes connect the global radiation budget to the weather. They range from large-scale structures, such as the circumpolar vortices that spawn the jet stream, down to familiar weather features like cold fronts and storms. So the radiation budget, while it is the reason for the existence of the weather, is not of immediate application to short-term forecasts. Nevertheless, it is crucial to any deep understanding of meteorology, to climate studies, and to the exploration of long-term weather fluctuations.

Satellites have provided values of the three radiation fluxes (incoming solar, upwelling reflected, and reradiated energy) that contribute to the earth's radiation budget, and they have done so for different geographical locations and seasons. The next step in this field will be to disentangle the detail: How much of the reradiated energy comes from the surface as opposed to the atmosphere, for example, and how absorption and reradiation vary with altitude and the physical composition of the atmosphere.

In the late 1950s, much controversy still surrounded the basic question of the earth's overall radiation budget. That controversy arose because scientists were calculating the radiation budget from too little data. They needed a Tycho Brahe, or several Brahes, who would collect the observations "heaped on observation," but before satellites, there was no practical way of collecting comprehensive data. Some sounding rockets and ground stations had measured values of, respectively, the incoming solar radiation and terrestrial radiation. And even before the administration announced the satellite program, the meteorology panel of the International Geophysical Year had plans to gather more ground-based observations to increase their understanding of the earth's radiation budget. Then came the satellite program, with its offer of a new perspective on the earth, that made possible Suomi's proposal.

Suomi and Parent did not join the International Geophysical Year until nearly a year after President Eisenhower's announcement that the United States would launch a satellite. Then, sometime in the spring of 1956, Suomi attended a lecture given by Joseph Kaplan, the "five cigar" chair of the U.S.'s National Committee—the man who had shepherded the U.S.'s IGY proposals past wary administration officials. Kaplan by now was promoting the IGY and the satellite program to a wider scientific audience and to the public.

As he listened to Kaplan's lecture, Suomi knew that he wanted to participate in the satellite program. His doctoral thesis—about ten pages of text and thirty of charts—had described the radiation budget of a cornfield. That is, he was interested in what happens to the sun's energy: what is absorbed, what reemitted, when and how. Such basic information is valuable to agriculture. And today such studies are also important for correlating satellite observations with what is actually happening on the ground.

Never niggardly with the scope of his ideas, Suomi thought he could study the earth's radiation budget from a satellite, and that the project would be "much more interesting than the radiation budget of a lousy cornfield."

After the lecture he spoke to Kaplan, who told him to contact Harry Wexler, then chief scientist of the Weather Bureau as well as the chair of the IGY's meteorology committee and a member of the U.S. National Committee of IGY. By now the satellite panel was assigning priority to the various payload proposals, and at this late stage the sponsorship that Wexler quickly provided was essential. "Harry rigged it up," said Suomi, "so my little proposal snuck in about three days after the door was closed. I got about $75,000, which in those days was an enormous amount of money."

Wexler and his colleague Sig Fritz worked on some of the theoretical aspects of the scientific consequences of Suomi's instrument and told the satellite panel that he and Fritz were interested in exploring whether data from Suomi's instruments would show a correlation between isolines of reflected radiation and large-scale weather. Wexler presented Suomi's ideas on June 7, 1956. During the next six months he also solicited the support of other influential meteorologists.

Although Suomi and Parent did not submit a formal proposal until the beginning of 1957, their new idea upset the proverbial apple cart. It was not the last time that Suomi would disturb nicely laid plans with a last-minute idea. He was to do the same with his idea for a spin-scan camera.

In the case of the radiation budget experiment, the overturned apple cart belonged to Bill Stroud of the Signal Corps of Engineers. His idea, like Suomi's, was too ambitious for the existing technology, but both were pushing the boundaries. And both saw conceptually the promise of the bird's-eye view.

Stroud wanted to photograph clouds from orbit. He had two broad purposes: first, to seek ways to identify cloud types from above, and second, to estimate the albedo of clouds.

The panel was interested in Stroud's idea but concerned about the difficulties of extracting any useful information from the photographs. This was the concern that Kellogg and Greenfield had expressed in their report of 1951 and that would surface again during the TIROS program. Nevertheless, by the time Wexler introduced Suomi's idea, the panelists had decided to back Stroud's work, though it was classified as one of the "priority B" experiments.

And that is the priority that the experiment retained until the turn of the year, when a flurry of events precipitated a face-off between Suomi and Stroud.

By the end of 1956, the panelists had money for only four experiments. One of these was to be the Naval Research Laboratory's satellite designed to measure temperature and pressure in space, another would be Van Allen's payload, the third was to be a geomagnetic experiment, and the fourth experiment was to be Suomi's.

Stroud must have objected, because when the panel next met early in 1957, they had decided that he and Suomi should compete for the fourth launch vehicle and that at the end of 1957 the country's leading meteorologists should decide between them. The two raced to prepare their satellites. But then *Sputnik I* was launched, followed shortly by *Sputnik II,* and President Eisenhower's administration brought in the Army. There were now enough launch vehicles, and all the meteorologists had to do was to endorse both project.

Stroud's payload was launched first, on *Vanguard II.* His camera worked, but its images were of little use because the satellite was precessing wildly. It was not possible to say where the camera was pointing. Stroud's problems sprang from the difficulties that arise with spinning bodies. The final stage of *Vanguard II* was spun so that it would be stable in flight. And the satellite was designed to spin up after separation from *Vanguard* so that it too would be stable in orbit.

When the IGY's satellite panel met for the last time to wind up business, on July 21, 1959, John Townsend, of the Naval Research Laboratory,

Verner Suomi posing with the Vanguard satellite that was launched on June 22, 1959. The satellite ended up in the ocean.

gave a report of the *Vanguard II* launch that gives a flavor of the difficulties and the unknowns of rocketry and satellites in those days. One theory was that there had been a collision between the third stage and the satellite. The idea stemmed from what they had learned about *Vanguard I,* the launch that generated the data for O'Keefe's deduction of the pear-shaped earth. In that case, the Vanguard team had asked Fred Whipple's group of optical trackers at the Smithsonian Astrophysical Observatory to locate the third stage for them. Whipple's group had found the rocket's final stage to

Robert Parent (left) and Verner Suomi work on the electronics of their radiation balance experiment.

be in a markedly different orbit from that of the satellite. "So," reported Townsend, "even after what appears to be a shutdown [of the third stage], there is enough propulsion in that unit as it cooks, or burps, to account for several hundred feet per second difference between the bottle [i.e., the third stage] and the satellite itself." In other words, there was enough difference in velocity to send the two into slightly different orbits, raising the possibility of collision.

The alternative explanation centered around the fact that the spin up and separation might have occurred at the same time. This could have

Explorer VII which carried Verner Suomi's and Robert Parent's radiation balance experiment into space on October 13, 1959. The white hemisphere attached to the mirror is one of the bolometers that gathered data for the experiment.

caused a problem because of the third stage's unburnt fuel. The bottle, Townsend told the group, was an unbalanced stovepipe with holes in it that was unpredictable mathematically. If the satellite's spin thrusters fired while still attached, the bottle would precess wildly and impart that precession to the satellite. Whichever explanation was correct, Stroud's experiment was in trouble because the satellite had no internal damping mechanism.

Six months later, on October 13, 1959, Suomi and Parent's payload reached space. Suomi had a receiver in his bedroom. He received four signals a day, two within 105 minutes of one another, then a similar set twelve hours later.

"We worked like dogs," said Suomi of his and Parent's work on the radiation balance experiment. Leo Skille, a lab technician who worked with Suomi and Parent, recalled that Suomi had an air mattress that leaked, and that after four hours Suomi would find himself awake on the hard floor. The mattress served as Suomi's alarm clock, and it was not uncommon for Skille to come in at eight in the morning and see both Suomi and Parent still working.

Suomi spent most of his time in electrical engineering because the meteorology department did not consider the satellite work an enormous blessing, although it did not hold him back.

Suomi and Parent amused observers as, sporting the bow ties favored by some academics, they strode around campus deep in discussion. When they encountered a grassy quad, Parent would—so the story goes—walk around it, while Suomi strode diagonally across.

"I'm a person," said Suomi, "who says 'lets do this.' It might be a crazy idea, but I don't let initial problems stop me. Bob was there as an evaluator. He didn't discourage me. He put a certain amount of reality into some of the wild things I thought about."

"Suomi would go off at tangents," said his technicians, "Parent would bring him back to Earth. Suomi would figure you could build an instrument as fast as he thought of it. We'd say, 'Watch out for Suomi on Monday mornings.' He's come in with a whole bunch of ideas he'd thought of over the weekend."

The engineers would wonder sometimes what he was getting at, and they wondered whether it was just they that didn't understand his idea or whether he was coming out of left field. He always wanted to know why something wouldn't work. "He could be stubborn," they said, "and there were big rows, but he never held a grudge."

Suomi and Parent's first satellite experiment comprised four ping-pong-sized balls on the end of Vanguard's antennas—hence its irreverent nickname. One was white and thus reflected visible radiation but absorbed the longer-wavelength radiation reflected from the earth. One was black and absorbed all wavelengths, and the other two were shaded in such a way as to discriminate between incoming and upwelling terrestrial radiation. Each contained a resistor that enabled a determination of the balls' temperature (the relationship between resistance and temperature for different types of material is well known). These numbers did not provide absolute values for the three different types of radiation, but they did show how the radiation balance varied.

Suomi and Parent did most of the early construction themselves. "We were probably as capable as any of the engineers," said Suomi. The two of them built four models: one for mechanical and vibration tests, one for electrical tests, one flight model, and one backup.

The bane of their lives were the transistors, which they knew comparatively little about because transistors were a new technology. Suomi would order them in batches of fifty, paying $35 for an npn type transistor and $65 for a pnp. (A $35 watch today, said Suomi, contains about five thousand transistors.) The transistors were supposed to work within certain temperature ranges, but most of them clustered at the extremes of the stated range rather than at the center, as became usual when the technology matured. They tested each component by cooking it in Paula Suomi's electric frying pan and then putting it in the freezer. Then they hooked each component up to an oscilloscope and checked its characteristics by observing the plots of voltage against time for various inputs. If it worked, it could be soldered onto a circuit board; if not, it went into the bin.

At each stage of the construction, there were laborious tests. "I can honestly say," recalled Suomi, "that we never took any short cuts. If we made any changes, we made the tests. We almost worshipped that rule.

"I remember distinctly that as we worked on something, we would work it to death, and if the improvement was not satisfactory, we didn't say we don't have to worry about that anymore. We'd go and find a blackboard and try to understand the problem at a deeper level than just something we were soldering. For example, we needed a power amplifier to work over a wide range of temperatures, and so we needed transistors that would work over a wide range of temperatures. We tried to use silicon, but the only thing then that could handle enough power was germanium, but germanium is temperature sensitive; therefore we put in resistors to modify the power input. The resistors had to be reliable at a range of temperatures, so we put in thermistors to compensate for different temperatures."

In the end, they had underestimated by a factor of a thousand the time needed for testing, but to Suomi's irritation, the flight model was not put through the severest tests. It was felt that testing wore out the payload.

Their task grew more complicated when the Army became involved. All of those assigned launches were asked whether their payload could be modified for Explorer. Suomi and Parent were among those who built payloads for both types of satellite and launch vehicle. Werner von Braun told Suomi that if he could not have his experiment ready by the launch date, the rocket would carry a plaque reading "this space was reserved for the University of Wisconsin."

Suomi and Parent's first launch attempt was atop *Vanguard* on June 22, 1959. That payload went into the ocean. When I asked whether he and Parent went for a drink after the launch failure, Suomi looked into the distance for a moment and said, "Yes, we went for a drink." He paused and added, "maybe more than one. We weren't thinking yet about Explorer. It hurt too much to lose the experiment on Vanguard. I could tell you a lot of things we did and said then, but they're unprintable."

But they had to think about Explorer very soon, perhaps even before their hangovers cleared, because the launch was scheduled for July 16. Suomi and Parent observed this second launch from the blockhouse at Cape Canaveral. Photographers and journalists only a quarter of a mile away watched the liftoff and saw with horror that the rocket appeared to head straight toward them. The photographers continued shooting as those around them dived under cars and trucks.

The rocket was, in fact, veering toward a populated part of Florida, and the range safety officer deliberately exploded it. The blazing rocket fell within 150 feet of the blockhouse. Suomi and Parent were trapped for more than an hour. When they were allowed out, they went to the smoldering rocket and hacksawed out their instrument package.

For the next launch, in October 1959, Suomi, who now saw himself as a jinx, stayed in Wisconsin. That rocket reached space, and the payload collected its data successfully.

The value of Suomi's experiment is not that it provided a global map of the radiation budget, because it did not. Rather it is in the existence of the idea and in the thought that went into exploiting this new place from which to make observations and in the painful lessons learned during the primitive days of the space age.

The first factor operating against Suomi was the low inclination of the orbit selected for *Vanguard* and later for *Explorer*. From the low inclination the satellite could "see" only a limited area of the surface and missed the poles completely. Secondly, the primitive state of computing and data processing imposed limitations. In June 1961, Suomi wrote to a science journalist that they had been drowned in data from *Explorer VII*. The satellite was supposed to have been turned off on October 13, 1960, but the timer failed, so it was still operating eighteen months later. Suomi wrote, "Records are now being corrected on a minimum scale—it's reassuring to compare records a year apart . . . the condition of the black sensor is nearly the same as originally, but the white sensor has discontinued due to exposure to the sun's radiation."

Though the results fell far short of the ambitious plan, the first attempt had been made to gather comprehensive data about the most fundamental process driving the earth's atmosphere and oceans. And Verner Suomi was hooked by the new field of satellite meteorology. He went on to fly radiation budget experiments on two of the TIROS series of spacecraft and on classified satellites. He later referred to results from all of these when he was arguing that the spin-scan camera, for which he is most well known, should fly on one of NASA's experimental spacecraft—the application technology satellite, or *ATS-1*.

Storm Patrol

People have told me I'm a wonderful salesman, but it took all of my salesmanship while I was in Washington [to persuade NASA to fly the spin-scan camera].

—Verner Suomi to author, May 27, 1992

The spin-scan camera was a giant step. It gave you a view you didn't have before.

—Robert White, former president of the National Academy of Engineering, to author, 1992.

In his long, narrow office at the University of Wisconsin, with awards hung on the walls (others are stuffed in drawers in the basement), Verner Suomi recalled the first spin-scan camera that he and Bob Parent proposed to NASA in the fall of 1964. "It was disgustingly simple. The stuff on the ground that you need to put the pictures together, that was not so simple."

The camera's job was to continually monitor the weather over one portion of the earth's surface.

The space-based elements of the idea were, indeed, conceptually straightforward: a spinning spacecraft in geostationary orbit,[1] a telescope, a camera, and a data link with the earth. The practicality was a little more difficult. "But," said Suomi, "one of the advantages was that we didn't know what the problems were, so they didn't hold us up."

Suomi and Parent's first proposal for a spin-scan camera, dated September 28, 1964, was a hastily thrown together three and a half pages of text and two pages of very simple diagrams. Parent was the electronics expert. Their proposal was called "Initial Technical Proposal for a 'Storm Patrol'

---

1. Satellites in an orbit co-planar with the equator at an altitude of nearly 22,300 statute miles are called geostationary satellites because they can be regarded as stationary with respect to a particular point on the Earth, because the satellite takes about the same time to complete its orbit as the Earth takes to turn once on its axis. The satellite, however, is not exactly stationary with respect to the Earth because of the gravitational abnormalities caused by the Earth's inhomogenous structure, but by the time of Suomi's and Parent's proposal, observations of satellites like Transit were beginning to improve the accuracy of gravitational models, making it possible to predict shifts in the satellite's position.

Meteorological Experiment on an ATS Spacecraft." Like other meteorologists at the time, Suomi wanted to take advantage of the 22,300-mile-high geostationary orbit, in which a satellite stays in the same position, more or less, with respect to the earth and thus "sees" the weather moving underneath. Polar-orbiting satellites, by contrast, see successive snapshots of the weather in different places as they move through their orbit.

A geostationary orbit, however, is a long way away from the earth, so Suomi and Parent described a telescopic camera that would enlarge the distant image. Since only a small part of the earth would fall within the field of view, some method was needed of scanning in the east–west and north–south directions to build up an image of the earth's surface. The satellite on which they hoped to mount the spin–scan camera would be spinning at a steady 100 rpm, and thus automatically would scan a line from east to west. After each revolution, the camera would shift its field of view slightly to build the full picture of the earth's disc. Over the years, several electronic and mechanical methods of achieving movement in the north–south direction were explored.

Suomi and Parent thought that the image could be built over ten minutes from one thousand scan lines, giving a resolution at the subsatellite point of six nautical miles. In their second, ten-page proposal to NASA a year later, the camera, which was designed cooperatively with the Santa Barbara Research Facility of the Hughes Aircraft Company, had an image built from two thousand lines and thus an improved spatial resolution.

Today's technological descendants of the first spin–scan camera scan sixteen thousand lines in thirty minutes. During severe storms they can build more frequent pictures of smaller regions. They observe in the infrared. Each radiometric reading is assigned a color, and a false-color image is created. From these images, meteorologists, infer wind speeds, which are particularly important for modeling atmospheric conditions in the tropics (within thirty degrees of latitude north and south of the equator), where the temperature differences are too small for satellite sounders to make distinctions.

Despite the greater spatial resolution of today's satellites, Suomi, talking in 1992, was not happy about the thirty-minute time interval between photographs. In his opinion, the ideal interval is the ten minutes that he and Parent first proposed in 1964 because in that time very little change in the weather and very little detail of an evolving weather pattern is lost.

In that first proposal, Suomi wrote, "The object of the experiment is to continuously monitor the weather motions over a large fraction of the earth's surface." He and Parent envisaged a camera that would observe the earth between fifty degrees of latitude north and south, which would, of course, encompass the meteorologically all-important region of the tropics.

Suomi quoted results from his radiation balance experiments on *Explorer VII* and several of the TIROS satellites to make his case, writing that the amount of radiation reflected from the tropics was lower than expected, even though the total outgoing radiation from the earth was close to earlier estimates. Thus, more heat than previously thought was being transferred from tropical to polar regions.

The questions meteorologists needed to answer were, How was that heat transfer achieved, and how did it affect global circulation of the atmosphere? They had few observations with which to work because the tropics—the "boiler," as Suomi wrote, of the giant atmospheric heat engine—which cover about half of the earth's surface, are eighty percent ocean. The polar orbiting TIROS satellites did not help much. Those satellites spent only about fifteen minutes traversing the region as they headed north (similarly for the southward journey) in their orbit. There was a gap of twelve hours before the spacecraft was next above the same subsatellite point. In the tropics, where weather patterns develop and dissipate in far less than twelve hours, the result was that the TIROS satellites did not provide observations of the complete life cycle of a typical tropical storm. Instead, meteorologists inferred the progress of a "model" storm from observations of different storms in different places at different stages of their development.

Yet these storms, including hurricanes, are one of the mechanisms by which the "boiler" of the "atmospheric heat engine" redistributes heat around the earth. The rationale of the spin-scan camera was to provide data that would allow meteorologists to explore these mechanisms.

It took more than a decade for meteorologists to find effective ways of exploiting the spin-scan camera, but eventually inferences of wind speeds in the tropics improved atmospheric models.

Bob Ohckers, an electronics technician who joined Suomi's group in 1967 from RCA, said that Suomi initially wanted to measure the winds from the displacement of clouds between successive images. "We'd get one image (an 8 by 8 transparency) in a frame and superimpose a second image taken twenty minutes later. First, we'd line up the geographical points in the two transparencies, completely ignoring the clouds. Next we'd shake the images in a frame until the clouds from the two images were superimposed on one another and the geographical features were displaced. You could tell when the clouds coincided because the light shining through from below was at it dimmest in those places. Then you would measure the x and y displacement of the clouds." The method worked, but it was impractical, and the department's software group came up with a better way of doing the same thing. When Suomi saw the results of the software, he dropped the mechanical approach without a backward glance.[2]

Although meteorologists in the early 1960s were keen to observe the earth from geostationary orbit and plans existed on paper for a geostationary meteorology satellite, there was a problem. "No one had any idea," recalled Suomi, "about how to get the blooming thing up there."

Then Harold Rosen, Donald Williams, and Tom Hudspeth, of the Hughes Aircraft Company, came up with the engineering concepts that made attaining geostationary orbit both economically and technically feasible at an earlier date than anyone had thought possible. It was an advance that was to be a key factor in opening up the multi-billion-dollar business of civilian communication satellites in the mid 1960s, but a description of a NASA satellite based on the Hughes design also fired Suomi's imagination. It was called, prosaically enough, the Application Technology Satellite-I. *ATS-I* was to carry an experimental communications payload with sufficient bandwidth to transmit a TV channel.

Suomi's attention was caught by the simplified block diagram that accompanied the article describing *ATS-1*. It looked to him as though the satellite should be able to carry a small camera and that there would be sufficient bandwidth to carry its images back to Earth.

During July and August 1964, Suomi elaborated his ideas, and he and Parent hastily put them into their September proposal to NASA.[2]

---

2. The idea of taking advantage of a spinning satellite for an automatic east-west scan was not uniquely Suomi's. A similar idea existed in the world of photo reconnaissance. Merton Davies, of the RAND Corporation, and Amron Katz had encouraged adaptation of a panoramic camera for high-altitude photography. Then Merton Davies had realized that the camera could be fixed to a spinning spacecraft to achieve an automatic east-west scan as a satellite, not in a geostationary orbit, traversed its orbit. From *RAND's Role in the Evolution of Balloon and Satellite Observation Systems and Related US Space Technology.* Edwin Land showed the first satellite reconnaissance photographs from such a system to President Eisenhower on August 25, 1960.

Earlier in the year, Suomi had completed a brief stint as chief scientist of the Weather Bureau, working for Robert White (who later became president of the National Academy of Engineering, retiring in 1995). "Wouldn't it be nice," Suomi now asked White, "to beat the Russians into space with a camera viewing the weather from a geostationary satellite?" Seven years into the space age, many space scientists and engineers still felt they needed to regain the technological initiative from the Soviets. White's practical response was to grease the bureaucratic wheels for Suomi, who, as with the International Geophysical Year, was making a belated entry into a satellite program.

NASA at first told Suomi that the spacecraft would not be stable enough for his camera. Suomi called Rosen at Hughes, who, incensed by the comment, made his own phone calls to NASA.

In the meantime, Suomi presented his and Parent's ideas to government officials and industry representatives, including TRW and the Santa Barbara Research Center of the Hughes Aircraft Company. Both companies invested their own resources to investigate the concept. Several data processing issues had to be solved. For example, the camera was being designed to have a precise geometry, and the geometry of the resulting image had to be preserved after processing. Second, from geostationary orbit, the Earth occupied only about 16 degrees of the camera's 360 degree field of view (because it was rotating). So the camera would be recording images of the earth for only about a twentieth of each revolution, and the signal would take up twenty times more bandwidth than was needed to relay the image data. There were questions, too, about the impact of camera distortion and about nutations of the spin axis (precession).

NASA backed the proposal in time for the camera to fly on the *ATS-1* spacecraft. Suomi kept the technical authority for the project at the University of Wisconsin but subcontracted the physical construction and final engineering to the Santa Barbara Research Facility.

Some years later, Hughes filed a patent on the spin-scan camera, but Suomi opposed them, supporting NASA's claim to the patent because it was the agency that had funded his work and because Suomi believed that the validity of the Hughes patent claim rested on his ideas. NASA, which would be less fortunate during a later patent dispute with Hughes about crucial elements of the Williams, Rosen, Hudspeth satellite design, won the dispute. Nevertheless, as Suomi said some years later, Hughes engineers made important contributions to the development of the camera, and, he added, ". . . Hughes built the camera, so in a manner of speaking, they reduced the idea to practice."

Suomi almost missed the launch of *ATS-1*. He had forgotten to do the paperwork for his security clearance, but a colleague interceded for him. Suomi said his most exciting professional moment came when the first image of Earth's disk ever taken from space appeared on an oscilloscope. The aim of the spin-scan camera had been to have weather imagery available to meteorologists in real time. That did not happen immediately. The first printed images from the spin-scan camera on *ATS-1* were ready four or five days after the launch. Suomi was scheduled to give a lecture at the American Meteorological Society. He said, "I had a whole bunch of negatives, and I tried to line these up with one another. I put a pin through, and I made a "movie." I gave my talk and ran the movie. They thought it was wonderful to see the clouds moving."

They had, in fact, seen the first ever animated picture of the earth's weather—the primitive precursor to the pictures that appear today on television weather forecasts. There was still a long way to go before the technology would be regarded as mature, but one of the two most significant classes of instrument (the other was the sounder) that would facilitate that process was aloft. And it was mounted on a satellite that was the technological kin of *Early Bird,* the world's first commercial communication satellite.

3

★

Communications

> **"The telephone network is the nervous system of our civilization, carrying messages of demand and direction, of pain and pleasure, to collective enterprises and to individuals alike."**
>
> John Robinson Pierce, director of research of electrical communications at the Bell Telephone Laboratories, writing in the *Atlantic Monthly* in December 1957.
>
> **"A true broadcast service giving constant field strength at all times over the whole globe would be invaluable, not to say indispensable, in a world society."**
>
> **AND**
>
> **"Many may consider the solution proposed in this discussion too far-fetched to be taken very seriously. Such an attitude is unreasonable, as everything envisaged here is a logical extension of developments in the past ten years—in particular, the perfection of the long-range rocket of which V2 was the prototype."**
>
> Extra-Terrestrial Relays: Can Rocket Station Give World-wide Radio Coverage? By Arthur C. Clarke in *Wireless World*, published in October 1945.

The Players

Direct evidence of field strength above the earth's atmosphere could be obtained by V2 rocket technique, and it is to be hoped that someone will do something about this soon as there must be quite a surplus stock somewhere.

—Extra terrestrial relays: can rocket stations give world-wide radio coverage, by Arthur C. Clarke. Published in *Wireless World*

On October 4, 1957, only thirty-six people in the United States could call Europe simultaneously, via AT&T's recently installed transatlantic submarine cable—TAT-1. If the ionosphere was stable that day, about a further one hundred high-frequency radio circuits would have been available.

AT&T laid TAT-1 in 1956. It was a power-hungry coaxial cable, costing $2 million. To provide enough bandwidth (a wide enough range of frequencies) for live television would have needed twenty such cables. It was not until the mid 1970s that live cable TV was theoretically possible. By then satellites already spanned the oceans, though submarine cables made from optical fibers would mount a stiff challenge to satellites in the 1980s. But that, as they say, is another story.

So, in 1957, two or three people per state could have called Europe at the same time; even fewer could have called countries on the Pacific rim; there was no live transoceanic TV; and the information superhighway was an idea beyond even the most exotic pipe dream. Though computers, television, and telephone all existed, the oceans were truly barriers to communication. And the world that contained these familiar-sounding technologies was very different from our own. After October 4, satellites, too, became a reality. Within decades, communications satellites had done much to change the world. Satellite communication is now a multibillion dollar business.

Where did the story begin?

Before there were spacecraft, there were science fiction writers. Most imagined that ground controllers would communicate via radio with their spacecraft. Then in 1945 a junior officer in the Royal Air Force spotted the unique advantages for communication of putting a satellite into an orbit where it maintained the same position with respect to its subsatellite point—geostationary orbit. The satellite would be like a huge microwave tower. Any antenna on Earth within sight of the satellite could beam a signal to it, which the satellite would then amplify before beaming it back to another antenna at a different spot on Earth.

For a satellite to seem to remain stationary, it must meet two conditions: the orbit must take the same time to complete as the Earth takes to rotate once around its axis (be geosynchronous), and the plane of its orbit must coincide with the plane through the equator (zero inclination). If a geosynchronous satellite has an inclination of zero degrees, it is geostationary, and its place in orbit is designated by the longitude of its subsatellite point on the equator. A satellite is travelling at the velocity needed—more or less—to maintain a geosynchronous orbit when it is at an altitude of about 22,300 statute miles. At such an altitude, the satellite is within site of nearly one third of the Earth's surface, excluding the poles. In 1959, satellites in this type of orbit were referred to as a 24-hour rather than geostationary satellites.

The junior officer in the RAF imagined that this orbiting telecommunications relay station would carry a crew, which, though wrong, was not a strange thought given the future science fiction career of the young man—Arthur C. Clarke. Clarke ran through some calculations, and in July 1945, he sent an article on the subject to the magazine *Wireless World*.

The editors were reluctant to publish something that seemed to them like science fiction, and they balked at acceptance. By October, they had relented, and the article appeared in print. It talked of field strengths and transmitter power, of solar power, and of how little time the satellite would spend in the shadow of the earth; and it suggested the best positions in orbit to provide a global system.[1] Clarke's predictions turned out to be prescient.

At the other side of the Atlantic, an electrical engineer at the Bell Telephone Laboratory, John Robinson Pierce, who knew nothing of Clarke's article, spent his leisure hours writing short science fiction stories and his working hours immersed in the complexities of microwave communication. Later, it was Pierce who was largely responsible for persuading NASA to carry out communication experiments with the passive Echo spacecraft in August 1960. Like the Moon, which reflected military communications between the East coast and Hawaii, Echo, which acted as a huge mirror in

---

1. In 1946, Louis Ridenour, an engineering professor at the University of Pennsylvania, independently put forward the idea that a satellite in synchronous orbit would be a good place for a radio relay.

the sky, bounced a signal across the U.S. The two-way Moon relay was operational between 1956 and 1962 and was manned when the Moon was in radio sight of both stations, usually for three to eight hours at a time. Often, when ionospheric storms shut down the usual radio channels, the Moon provided the only link to and from Hawaii for several hours at a time. In 1953, Pierce first suggested that if an artificial reflecting surface could be launched, it could bounce radio signals across oceans.[2]

Pierce joined Bell Laboratories after being awarded his doctorate by the California Institute of Technology in 1936. During World War II, he came across publications by an Austrian refugee, Rudi Kompfner, who was working for the British Admiralty. In 1943, Kompfner invented a class of vacuum tube, known as a traveling wave tube (TWT), that was to have an enormous impact on missile guidance and on communication through submarine cables and via satellite. In 1945, Pierce wrote his first paper about the new concept of traveling wave tubes and developed the first practical application of the technology. Kompfner would later say that he had invented the traveling wave tube but that Pierce had discovered it.

Pierce may as well be allowed to define a TWT in his own words, written in 1990 for a *Scientific American* publication, *Signals, the Science of Telecommunication*.

> The traveling wave tube is a type of vacuum tube that gives high gain over a broad band of frequencies. An electromagnetic signal wave travels along a spring-shaped coil of wire, or helix, while electrons in the high voltage beam travel through the helix at close to the speed of the signal wave. The electrons transfer power to the wave, which grows rapidly in power as it travels down the helix.

Essentially, the vacuum tube allows electrons to flow from cathode to anode with few collisions and permits an energy exchange from the electron beam to a radio wave constrained to travel the length of the tube. Thus the radio wave is amplified. There have been many versions of TWTs since 1945.

---

2. The U.S. Air Force had an alternative approach to passive communication. This involved distributing five hundred million copper threads with a thickness one-third that of a human hair into orbits two thousand miles above the Earth. These would be spaced at five hundred-foot intervals and create an artificial ionosphere. It was not a popular idea with optical and radio astronomers. The scheme was initially called Project Needles, but because the name seemed too descriptive, it was changed to Project Westford. An attempt to distribute the copper threads failed on October 21, 1961. In a research proposal prepared for internal consumption, John Pierce and his colleague Rudi Kompfner said that Project Westford had very little to recommend it.

Pierce was impressed by what he knew of Kompfner's work, and after the war he encouraged senior staff at Bell to recruit Kompfner. They were eventually successful, and Kompfner joined the lab in 1951. Pierce and Kompfner worked together cooperatively and productively for many years, and Kompfner was supportive of Pierce's interest in satellites. Their work was the starting point for the Bell team that designed a second type of communications satellite, one with an active repeater that would, like the satellites envisaged by Clarke, amplify the signal before radiating it to Earth. This satellite became known as *Telstar*.

Telstar was not a twenty-four-hour satellite but rather was planned for a medium-altitude orbit, and so could only be seen by two ground stations simultaneously for about twenty minutes. AT&T calculated that about forty 150-pound satellites in random medium-altitude orbits could provide a communications system with Europe. As soon as one satellite disappeared over the horizon, the transmitting and receiving antennas would lock on to the next mutually visible spacecraft. Such a system, said AT&T in the spring of 1961, could provide sixty channels by 1963 from North America to Europe and three thousand by 1980. The system would give ninety-nine percent probability of a satellite being simultaneously within sight of ground stations located in Maine and in Brittany.

In the summer of 1959, while the lab was still working on *Echo,* Pierce, Kompfner, and their colleagues at Bell were beginning to think that medium-altitude active repeaters rather than passive satellites were the most promising technology for transatlantic communication. For sound technical reasons, it seemed to them that geostationary satellites would not be feasible for many years.

On the West Coast, unbeknownst to Bell, a handful of engineers—Harold Rosen, Donald Williams, Tom Hudspeth and John Mendel—would soon solve, at least on paper, the problems then facing engineers considering a geostationary orbit. When Leroy Tillotson, at Bell, finished a technical paper on the specifications for a medium-altitude satellite in August 1959 and sent it to Pierce, Kompfner, and other senior members of the research department, Rosen and Williams were putting the finishing touches to their proposal for a twenty-four-hour satellite. The lightweight TWT designed by John Mendel, who had learned his trade in John Pierce's lab, was critical to the proposal. The proposal, in Rosen's and Williams's names, was the beginning of a development that led to *Syncom,* the first geosynchronous satellite, and to *Early Bird,* the first commercial communications satellite.

In Arthur C. Clarke's view, Pierce and Rosen are the fathers of communications satellites. During the early 1960s, however, there was little love lost between the two men. Rosen saw Pierce as obstructionist; Pierce thought that Rosen was making wild claims and would say anything to win support for his twenty-four-hour satellite. It is said they almost came to blows on stage during one conference. Yet they had far more in common than either could have imagined. Both were told at different times by their superiors that they could not go ahead with their work. "Cease and desist," is what Mervin Kelly, head of Bell Telephone Laboratories, told Pierce in 1958. Both had a fine disdain for the Department of Defense's plans for a twenty-four-hour satellite and for NASA's specifications for a medium-altitude active repeater called *Relay.* Both wanted to keep the government out of communications satellites.

Both, too, had been educated at Caltech, which in the 1930s was like an American Göttingen for the physicists and engineers who would become America's leaders in aerospace. Pierce was a contemporary of William Pickering and sought his cooperation for the *Echo* experiments. Rosen was Pickering's student and says that he was one of the teachers from whom he learned most. Each left Caltech in little doubt of his own intellectual ability.

The players, then, were John Pierce, Harold Rosen, and Donald Williams, with Tom Hudspeth, Rudi Kompfner, and John Mendel in strong supporting roles. Passive, medium-altitude active, and twenty-four-hour active satellites were the engineering concepts they contemplated.

Men and ideas fitted into a larger, more complex tapestry. It was not just that communications satellites were now within the state of the art, but there was also an increasing commercial and military demand for better communications.

TAT-1 remained the only transatlantic cable for telephony until AT&T laid a second link in 1959, bringing the number of simultaneous calls that cable could carry across the Atlantic to seventy-two. Adding these to the number of high-frequency radio circuits available on a good day, as many as four people per state could simultaneously have called Europe on the day of John Kennedy's inauguration as president in January 1961. It was still not possible to make live transoceanic TV broadcasts. Instead, recordings were flown by jet or fed slowly down cables.

In response to growth in demand, particularly for calls to and from Europe, AT&T planned to lay a third transatlantic cable in 1963, adding a further two hundred telephone circuits. Even this would not be enough to meet predicted growth in demand. But there seemed to be insurmountable engineering obstacles to developing higher-capacity cables, and the radio spectrum was already overcrowded. Worse still, solar minimum would occur between roughly 1962 and 1966. With less solar energy enveloping Earth, the ionosphere could be less active and thus would not reflect certain frequencies. Experts calculated that this would cut by two-thirds the available high-frequency radio channels worldwide.

The Department of Defense, with troops stationed around the world, often in places with which it was difficult to communicate, was deeply concerned.

With such a paucity of communications infrastructure coupled with the commercial and defense advantages of enhancing communications, it is not surprising that the Kennedy administration placed a high priority on the development of communications satellites. Communications satellites (and meteorology satellites) figured in Kennedy's famous moon speech of May 25, 1961.

And there were strategic advantages for the United States in developing communications satellites. Communications technology looked as though it could serve as a versatile foreign policy tool that could extend American influence throughout the world. John Rubel, deputy director of defense research and engineering (DDR&E) and for a while the acting director, pointed out in a white paper written in April 1961 that countries newly emerged from colonialism were often reliant for communication on their former colonial powers. He cited the cases of Guinea and Nigeria, which had to go through France and England to communicate with one another. It would be of "incalculable value" in the battle for men's minds, wrote Rubel, for the United States to maintain a lead in communications technology. "Many feel that the United States should support satellite-based telecommunications systems to achieve these aims, even though there were no immediate commercial advantages resulting therefrom."

The DDR&E held the third highest civilian position at the Pentagon, roughly on a par in some circumstances with the chairman of the Joint Chiefs of Staff. Thus Rubel was in a position of some influence. He had exerted that influence once, at the prompting of NASA administrator T. Keith Glennan, to change an agreement that NASA and the Defense

Department had made in November 1958, confining NASA to work on passive communications. It was an agreement that deeply frustrated NASA's engineers at the Goddard Space Flight Center. A new agreement, formalized in August 1960, permitted NASA to work on active communications satellites. Both parties observed a tacit understanding that NASA would work only on medium–altitude satellites, while Department of Defense developed twenty-four-hour satellites.

By April 1961, Rubel seems to have been feeling his way through a complex strategy that would also set aside this second agreement and permit NASA to develop Rosen's twenty-four-hour satellite. This was necessary because the Defense Department's own plans for a twenty-four-hour satellite, called *Advent,* were going disastrously wrong, but there would have been too much opposition to simply canceling the satellite and replacing it with Rosen's. If, however, the agreement between NASA and Defense could be set aside, then NASA could place a contract for the Rosen proposal. The agreement was dropped.

By August 1961, NASA had placed a sole-source contract with the Hughes Aircraft Company for a twenty-four-hour satellite, and the Department of Defense was to make the Advent ground stations available. The idea was that the twenty-four-hour proposal, now called Syncom, would be a cheap interim satellite to meet military needs until Advent could be developed. A year later, Advent was canceled.

The Syncom decision was a sweet triumph for Harold Rosen and Donald Williams. Before placing a contract for Syncom, both NASA and the Defense Department had been dismissive of Rosen's and Williams' engineering concepts. The first Syncom satellites were transferred to the military, and in the mid 1960s these provided links to Southeast Asia in support of America's growing presence in the region. Thirty years have passed, and Harold Rosen is not yet tired of telling people how the Army and Air Force rejected [his ideas] but within a few years had to rely on Syncom.

By funding the development of Syncom, launching *Telstar* (at AT&T's expense) and developing *Relay,* NASA enabled two approaches to a global satellite communications system to be demonstrated. When the International Telecommunication Satellite Organization (Intelsat) was formed in 1964, it was not yet clear whether international communications would be based on twenty-four-hour satellites or constellations of medium-altitude satellites. If one twenty-four-hour satellite operated successfully, however, Intelsat would know that it was well on the way to pro-viding a global system, whereas tens of Telstar (or *Relay*-like) satellites would have had to be launched to prove that a global communications sys-tem of medium-altitude satellites would work. Thus it was sensible to first launch one twenty-four-hour satellite, and the success of three Syncom satellites was encouraging.[3] The successful launch of *Early Bird* settled the question, and most commercial communications satellites today occupy geostationary orbits (the countries of the former Soviet Union use another orbit, one better suited to communications at high latitudes).[4]

Though the decision to "go geostationary" has been validated since 1965, the merits of the alternative technological approaches were still being debated in the early 1960s. The technical arguments were enmeshed in and obfuscated by a highly charged policy debate about the role of government versus private industry in the development of communications satellites.

The debate began during the closing months of the Eisenhower admin-istration, when T. Keith Glennan announced that NASA would facilitate the development of communications satellites by providing launch oppor-tunities for industry on a "cost reimbursable basis," which meant that industry would pay for the launch, but not at a true commercial rate. Glennan, like President Eisenhower, believed that private industry should be involved in the development of communications satellites. At the time, the most aggressive private industry in this field was AT&T. By October 1960 it had started Project TSX, which became Project Telstar, and had begun spending millions. Senior NASA staff and the attorney general were leery of AT&T. The company already had a virtual monopoly on voice transmissions. Neither NASA nor the Justice Department wanted to make decisions that would exclude from the new field companies that were not starting from the strong position of an existing monopoly.

When President Kennedy took office and James Webb replaced T. Keith Glennan, the emphasis shifted somewhat to a concern about how much control the government should retain over the development of communications satellites given their strategic importance. Webb, whom

---

3. Aerospace companies had their own ideas. Lockheed, for example, proposed that it, together with RCA and General Telephone and Electronics, should launch a system of spin-stabilized satellites into twenty-four-hour orbits. GTE had also had earlier discussions with the Hughes Aircraft Company.

4. The story today has come almost full circle, and commercial plans for fleets of medium altitude communication satellites pose a challenge to geostationary telecommuni-cation satellites.

AT&T viewed as anti-industry, said that he did not want to put AT&T up against the whole Soviet Union.

Engineers at Bell perceived that the debate had become truly heated in February 1961 when Lloyd Berkner, whose proposal of an International Geophysical Year ten years earlier had set so much in motion, said in a speech that communications satellites would be a billion-dollar business in ten to fifteen years. The newspapers picked up the comment. Congress took note, and the Justice Department quoted Berkner in submissions to the Federal Communications Commission and Congress. Berkner's comments were used to bolster the argument that space communication was too big for one company. Though his prediction was to prove correct, Berkner modified his views shortly after making them known, making the not unfamiliar claim that the media had exaggerated them.

But verbal arabesques could not change the course of the debate. Berkner had tapped into some widespread and deeply felt issues: the previous administration's concern about extending AT&T's monopoly; the current administration's desire to have some control over the development of a technology with strategic implications for the military, for commerce, and as a foreign policy tool; the current administration's concern that a private company should not represent the United States in negotiations for a global system; and industry's objection to being excluded by monopoly power from a potentially lucrative new market.

By February 1962, the Kennedy administration had sent a communications satellite bill to Congress. The bill set up a private company called Comsat under strict governmental control. Half the stock was offered to the general public and half to the common carriers. The Federal Communications Commission was responsible for distributing stock fairly between the common carriers, including AT&T. Key members of Congress had their own ideas about the bill, but by the end of the summer, the Senate had passed it, 66 to 11, and the House by 354 to 9.

President Kennedy signed the Comsat Act on August 31, 1962. It was the death knell for Telstar, though the concept of medium-altitude satellites had not yet been abandoned. Comsat would be the driving force behind the formation of Intelsat and thus behind the "go geostationary" decision. In the larger world of national and international policy, it was surely the right decision. To the engineers at Bell and to John Pierce, the man who pioneered the idea of commercial communications satellites and developed some of the critical technologies, the Comsat decision was a bitter disappointment.

Of Moons and Balloons

17

More than thirty years later, in his home in Palo Alto, John Pierce disposes concisely and precisely of questions about his pioneering days, tugging all the while at a bushy eyebrow. With his sloppy yellow Labrador retriever in attendance, Pierce reminisces politely about Echo and Telstar. Clearly, he has told the story many times, and he says, "I prefer to look forwards rather than back."

Asked to explain how a klystron works, he grows more animated. He becomes even more interested when writing down the names of mystery and science fiction writers he has not previously come across or talking of the Chinese poetry he translates, the haiku he writes, his admiration for Milton and Blake. Only when I asked him about the Comsat decision did passion flash with the sharpness of a disappointment almost, but not quite, forgotten. Pierce, a loyal son of "Ma Bell," would not leave Bell Labs, so the Comsat decision that excluded AT&T from international communication via satellites excluded him personally from a field he had pioneered.

As Pierce talks, his movements and speech are like those captured on video in the early 1960s, when he was the executive director of research at the Bell Telephone Laboratories. They are characteristically incisive movements, suggestive of someone who does not suffer fools or pretenders gladly. For a while, he thought that Harold Rosen was a pretender. "I was wrong," he says. He is less charitable about some of those he encountered at NASA headquarters during the Echo and Telstar days.

Pierce retired from Bell Labs in 1971 at sixty-one, an age when, as he says, he was still young enough to do something else. He joined the faculty of the California Institute of Technology for nine years, then moved briefly to the Jet Propulsion Laboratory as chief technologist. He is now visiting professor of music emeritus at Stanford, where he pursues an interest in the psychophysics of music—the relationship between acoustic stimulus and what we perceive internally. It is an interest he developed at Bell in his postsatellite days.

John Pierce's interest in science started when he was very young and his mother read to him from "very unsuitable books." Long before he could read, John could say words like electromotive force, even if he didn't quite know what they meant. "She was the mechanical member of our family," recalled Pierce. She also seems to have had faith in Pierce's mechanical ability, because when he and his friend Apollo built a glider, she went up with him, apparently unfazed by the earth flashing by beneath her feet. This, despite knowing that the first glider they had built had fallen apart as it taxied for takeoff. "I was crazy in those days," says Pierce, "doing things with very little information. I call it gadgeteering."

Pierce studied at Caltech, and after changing his major a few times, he settled for electrical engineering. He graduated in 1933, looked around at his Depression-era employment prospects, and decided he would be better off staying at Caltech. He gained his master's in 1934 and his doctorate in 1936. This time the world outside the ivory tower was less hostile to him, and Pierce got a job at Bell Laboratories.

He was told to work on vacuum tubes and left to get on with it, despite knowing next to nothing about the topic. This was typical of Bell Laboratories, where there was a lot of intellectual freedom to pursue research as well as the money to pay for it. Perhaps that accounts for the nobel prizes awarded to physicists at the lab.[1]

By the time World War II broke out, Pierce was expert in the basic theory and design of various classes of vacuum tube. He applied that knowledge during the war and learned a lot about electron optics and broadband amplification. Pierce contributed to the body of work that opened the spectrum above thirty megacycles, which before World War II was almost empty of artificial signals. Developing that technology was essential to the feasibility of communications satellites.

It was while undertaking a mathematical analysis of broadband amplification that Pierce came across Rudi Kompfner's work on traveling wave tubes. He was impressed. He wrote to Kompfner in 1946, adding his

---

1. John Pierce knew them all: Arno A. Penzias and Robert W. Wilson, who discovered the cosmic background radiation using the horn antenna developed for the Echo spacecraft; and John Bardeen, William Shockley and Walter H. Brattain, who invented the transistor. Transistor, incidentally, is Pierce's neologism. Walter Brattain asked him what to call the new device. Pierce writes in *Signals:* "I told him "transistors," it seemed logical enough. There were already Bell system devices called thermistors, whose resistance changed with temperature, and varistors, whose resistance changed with current. I was used to the ring of those names. Also, at the time we thought of the early point-contact transistor (then nameless) as the dual of the vacuum tube; in the operation of the two devices the roles of current and voltage were interchanged. The reasoning was simple. Vacuum tubes have transconductance, resistance is the dual of conductance, and transresistance would be the dual of transconductance, hence the name transistor."

voice personally to that of the management whom he had persuaded to recruit Kompfner.

After working on traveling wave tubes, Pierce and others at Bell Labs turned their attention to MASERs (Microwave Amplification by Stimulated Emission of Radiation). These devices generate or amplify microwaves. When they amplify a weak signal, they add little noise. It was the MASER at the heart of the ground antenna that made it possible to pick up the reflected signal from *Echo,* which was only a million-million-millionth of the ten kilowatt signal beamed to the satellite for reflection across the country. The MASER improved the antenna's sensitivity by a factor of one hundred compared with what Pierce had envisaged when he first speculated on the use of an Echo-like satellite for communication. And it was this MASER, protected from extraneous ground noise by a horn-shaped dish, with which Penzias and Wilson detected the cosmic background radiation.

So by 1954, many of the ideas and devices that were crucial for Bell's satellite communication work existed. And it was about now that Pierce became the first of the pioneers of communications satellites, which came about because he wrote science fiction stories (under the pseudonym J.J. Coupling, a concept familiar to electrical engineers). As a science fiction fan and author, Pierce was asked to give a talk on the subject of his choice to the Princeton, New Jersey, branch of the Institute of Radio Engineers. He must have had an erudite audience, given that RCA and Princeton University were nearby.

Over the years, Pierce had given talks about man in space, but he decided that for this audience he wanted a less fanciful subject. He began to wonder what role satellites could play in his own field of communication. At the time, says Pierce, communications satellites were "in the air," though it was a rarefied air. In 1952 he had written an article about interplanetary communication and had concluded that it was easier to communicate between the moon and Earth than across the United States. Now he did some quick calculations of the power needed for transmission to and from orbiting spacecraft and was surprised to discover that communications satellites were feasible.

Pierce gave the talk, which was to form the basis of his pioneering ideas for communication satellites.

Professor Martin Summerfield told Pierce that he should publish his talk. So, in November 1953, Pierce sent an article to *Jet Propulsion,* the journal of the American Rocket Society, which published it in April 1954.

The paper proposed three types of communications satellite: a one-hundred-foot sphere that could reflect a signal; a hundred-foot mirror in a twenty-four-hour orbit; and an active repeater in a twenty-four-hour orbit. The latter two, while theoretically stationary with respect to the ground, would actually be affected by solar and lunar gravity and so would need steerable ground antennas and stabilization by remote control.

The first of the three options—a hundred-foot sphere—was to be Pierce's inspiration for the Echo communication experiments in 1958. In 1954, shortly after Pierce's article was published, the U.S. Navy began experimenting with the voice transmissions to and from the moon that became the moon relay. But the moon is not an ideal reflective surface; its roughness gives multiple echoes at different wavelengths. A smoothly reflecting artificial satellite would, Pierce knew, provide a much higher-quality passive relay.

In 1954, few believed that satellites would be launched. Undeterred by the common view, Pierce told his audience in Princeton and wrote in his paper that if one found a way to build and launch a satellite, two classes of problem would remain, celestial mechanics and microwave communication. First, they would need to know where their satellite was and would be; then they would need to send and receive radio signals. All the satellite operators had to come to grips with celestial mechanics; some, like Transit, needed a very detailed understanding of the earth's gravitational field and its impact on an orbit. Pierce's paper acknowledged the problem but devoted more time to the issues of microwave communication: signal losses on passage from the satellite through the ionosphere and atmosphere to Earth (path losses); the diameter of transmitting and receiving antennas; signal frequency and strength; radio beam width; the method selected for superimposing the signal, such as voice or music, onto the radio carrier wave (modulation); the nature of the polarization of the radio beam; the frequency of the carrier radio wave; sources of noise (that is, other frequency sources that would make the signal difficult to hear); the power of the signal; the signal-to-noise ratio and the sensitivity of the receiver. These were among the topics that five hundred scientists and engineers would later address during the Telstar project.

The science fiction books that Pierce had begun reading as a teenager made spacecraft and radio communication commonplace ideas to him. So the topic of his talk to the Princeton radio engineers is not surprising. But his early love of science fiction also held him back. He had been so used to thinking of spacecraft as romantic fantasies that he did not at the time realize how close they were to realization in his own field of communication. Pierce discussed the idea of communications satellites with people around the lab, but he was concerned about the reliability of vacuum tubes (and who better to know their limitations) in space and the limited abilities of the primitive transistors that then existed. "I was conservative about satellites," he says.

Nonetheless, Pierce was responsible for persuading NASA to conduct communications experiments with *Echo*.

The satellite that became *Echo* was not initially intended to be a communications satellite. It was suggested by William O'Sullivan from the Langley Research Center when James Van Allen's satellite panel was selecting experiments for the International Geophysical Year. O'Sullivan wanted to launch a giant, aluminized Mylar balloon that could be inflated to a diameter of one hundred feet. With its small mass and large surface area, the balloons would be sensitive to comparatively small changes in force and thus would allow scientists to record how atmospheric density varied with position and solar activity and affected *Echo's* orbit. Van Allen's panel thought the balloon would be a good idea if they had sufficient resources for more than four launches.

In spring of 1958, Pierce and Kompfner read about the balloon and realized that it was exactly what Pierce had imagined would make an ideal passive communications satellite. They packed an ohmmeter and went to Langley to measure the conductivity of the plastic balloon. They decided that it would have a high reflectivity for microwaves. They took some samples of the aluminum-coated mylar back to the lab with them and confirmed its reflectivity. All they needed now was someone to launch the balloon. Unfortunately, the balloon was not one of the high-priority experiments for the IGY. NASA had not yet been formed, and the Department of Defense was already thinking in terms of the elaborate satellite that eventually became *Advent* and was to go so drastically wrong.

That summer, Pierce and Kompfner went to a meeting on communications satellites at Woods Hole. William Pickering (director of the Jet Propulsion Laboratory) was there and showed himself sympathetic to Pierce's ideas. Pickering suggested to the meeting that O'Sullivan's balloon would be ideal for a passive communication experiment. If Bell could find someone to launch the satellite, said Pickering, JPL's Goldstone ground station would participate in coast-to-coast communication experiments. To

Pierce, it seemed that Pickering's support was vital to the success of the lab's subsequent discussions with NASA.

Pierce returned from the meeting to a mixed reception. Mervin Kelly, the president of Bell Telephone Laboratories, asked a mathematician to study Pierce's proposal. The mathematician's report was negative. Kelly told Pierce to "cease and desist." Kompfner thought that their plans could go no further, but Pierce developed a severe case of deafness. He continued to think of Kelly as one of his heroes but concluded that "even great men" can be wrong. In October 1958, he delivered a paper on transoceanic communication via passive satellites to a national symposium on extended-range and space communications.

Later that same month, Pierce served as a consultant to the Advanced Research Projects Agency's ad hoc twenty-four-hour satellite committee. He listened to what he thought were impractical and inefficient proposals from "these completely uninformed men." It was clear after this meeting, in which elaborate satellites were discussed, that the Department of Defense was not going to launch the hundred-foot balloon. And shortly afterwards the Defense Department and the newly formed NASA agreed that the Department of Defense would develop active satellites and NASA would develop passive ones.

In the meantime, William Pickering had remained interested in a communication experiment, and NASA had been born. The new agency immediately inherited the Langley Research Center and the Jet Propulsion Laboratory.

In November 1958, T. Keith Glennan, NASA's administrator; Hugh Dryden, the deputy administrator; and Abe Silverstein, the director of space flight development visited Bell. The purpose of the meeting was to discuss global communications problems. Pierce made a general presentation about communications and satellites, and Kompfner talked about components, data processing, and tracking and guidance philosophy. The NASA contingent was interested, but nothing seems to have come of the discussions.

At the end of December, NASA took over Project Vanguard, thus gaining control of a launch vehicle. By January, the agency was showing an interest in BTL's ideas. On January 22, NASA, JPL, and Bell discussed what they hoped to learn from transmitting a signal between the East and West Coasts. Kompfner wrote to Leonard Jaffe, who headed NASA's communication satellite work, on February 10, urging him to let Bell know soon whether the project with JPL was to go ahead because of the large amount of work that had to be done. Less than a week later Kompfner warned all technical staff that a considerable amount of work of an unusual nature was coming up. Until now, Pierce had been deeply involved in selling the project to NASA. Now he took a back seat. The day-to-day running of the satellite work was handed over to Bill Jakes, who was responsible to Kompfner. At the end of February, Jakes was immersed in technical discussions with NASA about the MASER and how much bandwidth was needed given the signal and its Doppler shift. Pierce was already toying with the idea of active broadband satellite communication. NASA was by now enthusiastic and was contributing more money to *Echo*s than was Bell. The lab was building the horn antenna for experiments.

On June 10, 1959, there was a large meeting of all those involved with the project. O'Sullivan reported that one full-size balloon, including its inflating mechanism, was already being tested.

The balloon was made of 0.0005-inch-thick Mylar, coated by a 2,000 angstrom layer of aluminum. It weighed 136 pounds and had an optical reflectivity of seventy-five percent for tracking and a radio reflectivity of ninety-eight percent. Being passive, it did not have the complicated electronics needed for active repeaters, but it did carry a radio beacon so that it could be tracked by Minitrack. The sphere was to be inflated in one second by the release of four pounds of water through a plastic nozzle in the sphere. Langley calculated that the vapor pressure would last for seven days, and they were testing six subliming solids in an effort to extend the lifetime. After seven days, they expected a gradual loss of pressure because of micrometeorite impacts. The sphere would get wrinkled, decreasing its usefulness as a reflector for communications. Between November 1959 and July 1960, BTL and JPL practiced bouncing signals to one another first off the moon and then, three times, off *TIROS*. Their pointing accuracy needed to be good because *Echo*'s dimensions at an altitude of 1,000 nautical miles would be the equivalent of an object a little over an inch long a mile away, and it would be moving at four and a half miles per second.

The first launch attempt failed because the balloon did not inflate. But the second attempt, on August 12, 1960, was a success. Tracking *Echo* turned out to be tricky. The original plan was that NASA, at Goddard, would compute the orbital parameters and turn them into tracking instructions for Goldstone and the Bell antenna. A.C. Dickieson, Transit's project manager, writes in an unpublished manuscript that the tapes as received were late and full of errors. More success, he says, was achieved by taking orbital parameters generated by the Smithsonian Astrophysical Observatory and calculating tracking errors locally. Errors in orbital prediction were, however, inevitable in the fall of 1960. Only a few months earlier Bill Guier had predicted *Transit*'s position in orbit and realized how much more complicated the earth's gravitational field was going to be than anyone had thought.

From Bell's perspective, *Echo* provided background information for system planning and the design of Earth stations—information that fed into the Telstar project. *Echo* also demonstrated the effectiveness of the lab's low-noise receiving equipment as well as the predictability and stability of the transmission path.

Echo was the first satellite that was visible to the naked eye, and T. Keith Glennan had anticipated that it would cause a sensation. It did, bolstered by AT&T's brilliantly executed publicity campaign. On the night of the launch, the company sponsored a news special on NBC. It was replete with portentous music and massive radio telescopes. Another AT&T-sponsored video opened with 'America the Beautiful' and tugged at the patriotic heartstrings. AT&T won the publicity stakes hands down, but alienated NASA. The company's expropriation of *Echo* did not win it any friends at the agency. Pierce wonders whether AT&T's publicity success with *Echo* influenced NASA's selection of RCA for the *Relay* satellite. If it did, it was a minor influence compared with the much larger policy issues that were at stake.

Echo's success, technically and with the public, encouraged AT&T to go ahead with the development of a medium-altitude satellite as a prototype of a global system of communications satellites. That satellite—*Telstar*—became every bit as famous as its predecessor.

Telstar

Telstar captured the popular imagination in a way that it is hard to believe any satellite, especially a communications satellite, could do today. Perhaps it was the name; euphonious enough to make the satellite the eponymous star of its own pop song, released by the British pop group the Tornados. Certainly AT&T's impressive publicity machine, one that rivaled even that of NASA, aided the process.

Telstar was launched at 8:35 GMT on July 10, 1962. According to AT&T, more than half the population of the U.K. watched its first transatlantic transmission, a remarkable percentage given that far fewer people than today owned television sets. Only two Telstar satellites were ever launched, because even before the first went into orbit it was clear that Comsat, not AT&T, would be responsible for operating international communications satellites.

Instigating the Telstar project was one of the boldest moves ever made by a private company. Fred Kappel, chairman of AT&T, said that the company would spend $170 million—a considerable sum for the time—on an international communications system if the government would either step aside or facilitate development.

Project Telstar had five objectives: to test broadband communication, to test the reliability of electronic components under the stress of launch into space, to measure radiation levels (electronic components fail and data are lost if radiation alters the dopants in semiconductor material), to provide information on tracking, and to provide a test for the ground station equipment.[1]

A.C. Dickieson was appointed to head Telstar development in the fall of 1960. In his unpublished manuscript he says that he was told to go ahead in the shortest possible time and that he had whatever power and authority he needed to do his job. The team immediately started "spend-

---

1. Bandwidth measures the frequency spread. John Rubel's figures in a memo from DDR&E were that 20 words per minute in Morse code takes 9 cycles, 100 words per minute by teletype channel needs 75 cycles, voice telephone takes 3,500 cycles (3.5 kc), scrambled voice takes 50,000 cycles (50 kc), commercial TV takes 6,000 cycles (6 mc). Today we would say Hertz rather than cycles.

ing with gusto." For a while Dickieson did not know to the nearest $5 million just how much money was pouring into the project. But by early in 1961 he had spending under control.

The project's starting point was the early work by Pierce and Kompfner.

On January 6, 1959, while Bell and NASA were still discussing whether they would collaborate on a communication experiment with a passive satellite, John Pierce was arguing internally that Bell should assume a leading role in research in satellite communication, because it seemed likely that "satellites will provide very broad band transoceanic communication more cheaply than submarine cables." He wrote, "Active repeaters in twenty-four-hour equatorial orbits stationary above one point of the earth have many potential advantages but pose severe problems of launching, orientation, and life, whose solutions lie beyond the present state of the art." On the other hand, passive satellites were within the state of the art. Bell could, argued Pierce, undertake a program of research because more knowledge about low-noise MASERs and horn antennas would be valuable irrespective of the orbit that communications satellites would eventually occupy. There was so much activity in the field, wrote Pierce, and so little was known.

A few months later, Pierce and Kompfner were contemplating medium-altitude-active rather than twenty-four-hour satellites. There seemed to be just too many technical obstructions to developing the latter.

To John Pierce and his colleagues at Bell, the greatest drawback to twenty-four-hour satellites was the six-tenths of a second that would pass between one person speaking and the other person hearing what was said. The listener, thinking that a pause meant the speaker had finished, might then interrupt. Nothing could be done about the delay because that is how long it takes a signal to travel to and from a satellite in a twenty-four-hour orbit. Also, at that time, the equipment that suppressed echoes from the far end of the telephone line was not very effective, and Pierce was concerned that poor echo suppression coupled with the time delay would make communications via twenty-four-hour satellite intolerable. If, that is, one could have placed a satellite in a twenty-four-hour orbit in the first place. The launch vehicles and guidance and control systems needed to place a satellite in such an orbit were only then being developed.

Power was another problem. Imagine that power is distributed over the surface of a sphere and that the greater the distance of the receiver from the antenna, the greater the surface of the sphere over which the same power is distributed. Thus, if the distance increases by a factor of two, the power available at a particular point will be reduced by considerably more than twice—the *square* of the distance between transmitter and receiver.[2]

The thinking was that if one was to receive a usable radio signal on Earth from a satellite 22,300 miles away, one needed a high-gain, directional antenna on the spacecraft that would not waste power by broadcasting needlessly into space (high-gain directional antennas are common today). But if the satellite was to carry a directional antenna, then the satellite's orientation with respect to the earth—its attitude—would have to be controlled so that it remained constant relative to the earth. This called for a more complex, weightier design. Even at the best of times, satellite designers do not like to add weight. When no launch vehicle yet exists that can reach the desired altitude, the potential weight of a satellite is an even greater problem.

And then there was the all-important question of station keeping, which is essential for a satellite in a twenty-four-hour orbit. The satellite must maintain its position in orbit relative to the subsatellite point despite radiation pressure, lunisolar gravity, and inhomogeneities in the earth's gravitational field. If the twenty-four-hour satellite was to be constantly accessible to many ground stations and its antennas were to stay pointing toward those ground stations, then some mechanism for station keeping was needed as well as the fuel to operate that mechanism: again, extra weight.

Finally, launch vehicles do not place satellites directly into a twenty-four-hour orbit. The satellite is injected first into an eccentric orbit with its perigee at the altitude at which it separates from the launch vehicle and its apogee at geosynchronous altitude. So a twenty-four-hour satellite would need a motor that could fire at apogee and circularize the orbit: yet more weight.

2. How clearly a signal is heard depends then on the power of the transmitter and the distance over which the signal is sent as well as on the signal-to-noise ratio and the bandwidth of the transmission. The noise, heard as static, comes from many sources. It might, for example, be interference from terrestrial transmissions, such as the Baltimore TV station that drowned out *Transit 1A*'s signal or the electromagnetic field generated by vibrating electrons in the receiving circuitry.

Pierce and Kompfner were aware of all these problems and the possible poor quality of communication via a twenty-four-hour satellite. Further, Pierce, who from the beginning had participated in the panels planning military satellites, was aware of and unimpressed by the Advanced Project Research Agency's (later the Army's) plans for a twenty-four-hour satellite.

So, in March 1959, when Pierce and Kompfner wrote a paper demonstrating the theoretical feasibility of active broadband satellite communication, it was a constellation of medium–altitude satellites that they had in mind, not twenty-four-hour satellites. Bell still intended to work first with passive satellites simply because they would be ready first, but, wrote Pierce, the research department would work towards simple active repeaters. "If we manage to make better components than others and if we make a sensible design, we might take the lead in this field with a comparatively modest expenditure of money and effort. The tendency of ARPA has been to project elaborate and complicated schemes . . . our own course has been to get into actual experimental work with NASA and the Jet Propulsion Laboratory as early as possible. In this way we will encounter in an experimental way certain problems such as large antenna, low-noise receivers, tracking, modulation systems, orbit computations and the reliability of components which we believe will be important in all satellite communication systems."

Even though active satellites were at the forefront of Pierce's mind from the spring of 1959, he was publicly cautious, writing in June of that year to Hugh Dryden, the deputy administrator of NASA, "Right now, we don't feel we are able to evaluate the merits of active satellite systems as compared with passive systems well enough to propose any concrete steps as the next desirable thing to do."

Bell's efforts in the field of active repeaters began to solidify when Leroy Tillotson completed a major memo on active satellite repeaters on August 24, 1959. Much of what he described was similar to what would, after the launch of *Echo* in August 1960, become an internal Bell project, labeled TSX, which was later renamed Project Telstar.

Central to Tillotson's plan was the development of a six-gigahertz traveling wave tube. Pierce and Kompfner requested more information and were briefed in October and November 1959. An active-satellite planning committee was formed. Kompfner and Tillotson, but not Pierce, who was more senior, were members. They continued to meet until the research effort on active repeaters turned into a full-fledged development project in the fall of 1960.

At the beginning of 1960, the work on *Echo* was proceeding well and the research staff were considering what to do next. One possibility was a larger passive satellite with enough antenna gain (on the ground), bandwidth, and transmitter power to relay television signals across the Atlantic. An internal ad hoc group put together an outline of such an experimental system for NASA. As it became apparent what kind of technical pirouettes would be needed for the successful transmission of even the lowest quality TV signal, those working on the proposal became convinced that active satellites were needed.

Thus consensus emerged at Bell that the main effort should be on active satellites—along the lines suggested by Tillotson. A memo from Pierce to Jaffe says that Bell would be putting a major effort into active satellites during the next couple of years. Another letter from Kompfner to Jaffe reviews Bell's research on a long-life traveling wave tube and on the effects of radiation damage to solar cells and microwave circuitry. This work continued at a modest level in Bell's research department until after the successful launch of *Echo* on August 12, 1960.

News of Bell's interest in medium-altitude satellites did not filter out widely until the spring of 1960. When it did, some speculated that the passive scheme had been a smoke screen to cover Bell's real intentions. If it was, it had not been embarked upon as a smoke screen, though perhaps it was allowed to become one.

While Bell's scientists and engineers pursued theoretical calculations, AT&T's management in New York had been working on policy issues. In July 1960, AT&T argued before the Federal Communications Commission that frequencies should be reserved for satellite communication because the company was convinced that satellites would be more economical than submarine cables for transoceanic communication. A filing on July 8 disclosed AT&T's plans for a global communication satellite system costing $170 million. This plan called for fifty satellites without attitude control in a three-thousand-mile polar orbit. During the next twelve months the number of satellites, their altitude, and their design would change, but the idea of medium-altitude orbits remained.

In the midst of its political and technical preparations for an active satellite system, on August 9, 1960, the Department of Defense finally released NASA from its agreement to develop only passive satellites.

AT&T now focussed on persuading NASA to select its ideas for an agency led project to develop active communication satellites. On August 11, Pierce and colleagues were briefing senior NASA staff at headquarters in Washington on Bell's medium-altitude active repeater work. They told NASA of AT&T's discussions with the communication administrations of Britain, France, and Germany, and of those countries' interest in joining AT&T in satellite communication experiments. Bell's idea was that the Bell System would pay all costs, except those of general interest to the space community, such as investigating radiation effects. There was some discussion, too, about the technical difficulties of providing two-way channels (with *Echo,* voice went via satellite one way and came back over a terrestrial link).

The day after that meeting, *Echo* was launched. It was the brightest object in the night sky. In Ceylon, as Sri Lanka was then called, Arthur C. Clarke looked upwards and followed its passage with wonder. In the United States, AT&T's highly efficient publicity machine ensured that when the public gazed upwards, it was AT&T's name rather than NASA's that sprang to mind.

AT&T's publicity capsized discussions with NASA about launching a satellite based on AT&T's ideas. A little over two months after Echo's launch, in a meeting about Bell's plans for transoceanic communication via active satellite, T. Keith Glennan told senior staff from Bell and AT&T that the company's methods of publicizing its plans, including making misleading statements, had created difficulties for NASA. Glennan's remarks about publicity were only a small portion of the criticisms he made. He said that AT&T was not taking account of the "facts of life" and acknowledging through its planning the limited availability of launch vehicles, the problems of scheduling launches, and the aims of NASA's research and development program.

By the time of this meeting, on October 27, 1960, NASA's plans were known publicly to include medium-altitude active repeaters, and AT&T was now one of several companies waiting for the agency's formal announcement of a competition. Whichever company won, NASA also intended to further the development of communications satellites through "cost reimbursable launch support for private industry."

So, despite Glennan's stern admonishments in October 1960, the auguries were not entirely unfavorable for AT&T. NASA, under the leadership of T. Keith Glennan, favored the involvement of private industry in the development of communication satellites. AT&T had a vibrant research effort in the supporting technology for medium–altitude active repeaters, and *Echo* had been a spectacular success.

The days of the Eisenhower administration, however, were numbered. Transition to the Kennedy administration would soon be underway. Though both administrations were concerned about the antitrust implications of policies that favored AT&T, some in the Kennedy administration, notably James Webb, were additionally concerned about the strategic implications of communications satellites and thought that government should retain control over their development. The difference between the two administrations was apparent in a small but telling difference in the budgets that each submitted to Congress for FY62. Eisenhower called for private industry to contribute $10 million toward NASA's communication satellite program. Kennedy's budget allocated that $10 million from the public coffers.

This was the first faint stirring of the policy upheavals that would exclude AT&T from providing international satellite communications, a role that the company saw as its own as a matter of public trust. In the end, AT&T would have to be content with being one of the common carriers owning Comsat stock.

In January 1961, when John Kennedy was inaugurated, and for several more months, AT&T moved confidently forward with its plans. Bell Labs was working on ways to keep the satellites' weight down by developing sensitive ground-based antennas that would allow the satellites to transmit at as low a power as possible. That month NASA issued a request for proposals for a medium–altitude satellite to be known as *Relay.* During the *Relay* competition, AT&T and NASA suspended discussions about the possibility of the agency launching the telephone company's satellites. Bell, along with six other competitors, prepared a proposal. The others included RCA, which eventually won, and an outsider with no experience manufacturing satellites—the Hughes Aircraft Company.

Bell, which did not think much of NASA's technical specifications, submitted three versions of its proposals. One matched what NASA wanted, including frequencies and a radiation experiment as specified by the agency. The second retained NASA's radiation experiment but worked at the frequencies that Bell (AT&T was discussing frequency allocation with its overseas partners) thought would eventually be selected for an operational system (AT&T and Bell were correct). The third proposal was Bell's own design.

On May 18, NASA announced it was awarding the contract to RCA. At the technical debriefing, NASA told Bell that it was the best of the "amateurs" to submit a proposal. Bell's weaknesses in its bid, according to NASA, were many. The agency awarded the lab poor marks for solar cells made from n-on-p semiconductors rather than p-on-n. Yet Bell's scientists knew that the Evans Signal Laboratory had found that n-on-p semiconductors were more resistant to proton and electron bombardment than p-on-n semiconductors. Bell had fabricated n-on-p solar cells in December 1960 and confirmed the finding (these are what are used now).

The Bell proposal got a particularly low score for the low power of its transmitters, which would call for a low-noise ground antenna, and particularly for further improvement of the MASER that was used for Echo. The agency considered that Bell's tough specifications for a low-noise ground antenna would make the ground stations that NASA and others might build marginal. Its own course of action, said the agency, did not press the state of the art quite as hard. NASA judged, however, that Bell's traveling-wave-tube design was excellent and that its radiation experiment was good. The agency asked Bell to design the radiation experiment for *Relay*. Bell's view of this critique was that only one criticism—about a VHF antenna—was valid.

Preparing the stack of printed material for its proposal, wrote Dickieson, cost several hundred thousand dollars and "chewed up the time of a lot of key people who were sorely needed in designing the company's own satellite and ground station. But we really had no choice but to bid; without launch support, all of our work would be wasted, and NASA controlled the launching."

To the engineers at Bell, NASA's announcement that RCA had won the contract meant that they could revert to their own ideas for what would be known as Project Telstar. The announcement also left them with one rather major difficulty—the matter of a launch vehicle. Pierce and Kompfner joked with a visiting Soviet scientist that perhaps his government would launch the satellite.

A more practical discussion was going on between James Webb and Fred Kappel. Webb had called Kappel on May 18 to notify him that RCA had won the contract for *Relay* and to say that NASA was prepared to launch AT&T's satellite. Kappel responded that it was important for AT&T to go ahead with its satellite plans. Hard negotiations, led by Webb and Dryden, followed. NASA insisted on being reimbursed for launching the two satellites and that AT&T should assign any patents resulting from the development to NASA. AT&T agreed, and *Telstar,* which Wilbur L. Pritchard later said was superbly engineered, was underway.

An important aspect of the Telstar project was the two ground stations that Bell Laboratories built. The sophistication and sensitivity of these stations enabled Bell to put less power on the satellite. One station was in Andover, Maine, and the other in France. Britain used its own existing antenna but was unable to receive signals on the first night because of an unfortunate misunderstanding about the polarization of the signal.

The ground station in Maine was 170 feet long and three stories high, and rotated to follow the satellite from horizon to horizon. It was intended to be part of the eventual operational system and had to work in any weather, so Bell contracted for a radome (a radio transparent, domelike shell) to protect the antenna. The wall to which the radome was to be fixed was a massive structure. Someone quipped that in a thousand years, scientists would debate why the entrance door was in precisely that place. Excavation for the antenna foundations began in May 1961, and in January 1962, the antenna was complete. A temporary radome was erected until the special air-supported fabric of the final structure could be delivered. It sagged under the New England snow. Efforts to dislodge the snow failed until someone took a shotgun from the trunk of his car and shot holes in the temporary cover.

Testing of *Telstar* began in November 1961 and took 2,300 hours. The satellite was due at Cape Canaveral in May 1962 but was delayed for two weeks until a loose wire could be traced. The launch was set for July 10. In late May Bell heard about Project Starfish, a high-altitude atomic-bomb test. They were worried that if *Telstar* was in orbit, radiation from the explosion would seriously damage its electronics. They relaxed when they heard that the explosion was set for the day before their launch, believing that by the time *Telstar* reached orbit, the worst would be over. They later learned, as did APL, which had its TRAAC satellite aloft, that fallout persisted and precipitated along magnetic field lines. Telstar began to falter on November 18, 1962, and failed the following February.

Nonetheless, Telstar allowed an examination of the signal after passage at various angles through the ionosphere, the earth's magnetic field, and the atmosphere. Different methods of frequency modulation were tested, and the impact on available bandwidth assessed. These and other results were published in the open literature. Harold Rosen, of the Hughes

Aircraft Company, said, "Telstar showed that there were no propagation anomalies, that it was easy to calculate what the propagation would be like. It was a confidence builder."

Telstar attained orbit despite antagonism and suspicion between AT&T and Bell on the one hand and NASA on the other. Exactly how these tensions, which also existed under T. Keith Glennan, played out in the decision to select RCA for Project Relay is not easy to discern. Robert Seaman, who was NASA's associate administrator at the time, said in his exit interview that the message came through loud and clear from the Kennedy administration that the AT&T design was not the one to pick. In 1966, Webb said that the RCA proposal was clearly the best proposal for the research requirements of NASA, even if it was ". . . not necessarily the best as the first step towards an operational satellite system as desired by AT&T." Webb's phrasing is telling.

The disagreements between NASA and AT&T covered everything from choice of frequency to operation of the ground stations and negotiations with foreign telecommunications companies. AT&T was particularly jealous of the relationships it had built over the years with common carriers in other countries. The transatlantic submarine cable *TAT-1,* for example, was a joint venture with the British and Canadians. A British cable ship had laid the cable.

It is also clear from Dickieson's unpublished manuscript in the AT&T archives that Bell's technical people did not respect the technical ability of some of those with whom they dealt at NASA and that they thought much of the required paperwork pointless. Dickieson wrote, "the NASA people assigned to receive this paper were interested in the shadow, and not the substance, so we were able to keep them happy without [having them] interfere with our work."

The attitude of the Bell engineers comes through best in this following anecdote related by Dickieson. The National Physics Laboratory in the U.K. wanted to use *Telstar* in an experiment with the U.S. Naval Observatory to synchronize clocks in the two countries (this was before atomic clocks). Dickieson set things up. The experiment was performed and the results published. "At a subsequent meeting, Leonard Jaffe brought up the subject, and made it clear that the approved method was: first, discussions between the state departments of the two countries; second, reference to the technical organizations of the two governments, and finally down to scheduling by the Ground Station Committee." Says Dickieson, "I did not argue the matter, because I thought that if another useful experiment appeared, we would do it first and argue later."

Despite all this, the two organizations successfully launched the two Telstar satellites.

From February 1962, when President Kennedy submitted his Comsat bill to Congress, it was clear to Bell's engineers that the lab and AT&T were out of international satellite communication. They still had both satellites to launch, and the team worked on, fueled by the need to prove that private enterprise could operate in the field. To the public, the battles behind *Telstar* were unimportant. To them *Telstar* was the satellite that first broadcast live transatlantic television and promised a new era of international communication. Among those watching were a group of engineers at the Hughes Aircraft Company, in Culver City, California. After rejections and ridicule, they had won a contract for their ideas for a twenty-four-hour satellite in August 1961—nearly a year before *Telstar* went into orbit. The engineers watching *Telstar*'s broadcast were envious. They had dispatched a telegram of congratulations to Bell but were eager to see their own satellite in orbit. One of them, Harold Rosen, said "It was interesting in two respects, one was the beautiful picture coming from overseas. And two, it didn't last very long." They knew their satellite would be altogether different.

19 The Whippersnapper

I remember that when we were working towards *Telstar*, Harold Rosen and some colleagues came to visit. He was arguing for a geosynchronous satellite and putting forward every reason he could think of. I thought he was a whippersnapper, that he was just saying anything he could to get support.... He turned out to be an inspired electrical and mechanical designer.

—John Pierce to author,
speaking about Harold Rosen on October 2, 1995.

Looking back, you have to admire them.

—Robert Davies, chief scientist at Ford Aerospace,
formerly with Philco, in an interview with the author.

Pat Hyland is one of those people who are referred to as larger than life. He died in 1992 at the age of 95, having lived in a world where gals were gals and alcohol had yet to be banished from the corporate board-room. In his time, he made some spectacular mistakes, succeeded spectacularly, and knew and influenced the "great and the good." By 1958, he was running (and rescuing) the Hughes Aircraft Company, operating, so he recalls, an open-door policy through which any of his staff could walk. Through that door, one morning, walked Harold Rosen and Donald Williams.

This is how Hyland recalled the story in 1989.

He had known that Rosen and Williams, who were comparatively junior engineers, had some "harebrained" scheme for putting up a satellite, and that they thought it was pretty good but couldn't get anyone to sponsor it, and that it was going to cost a lot of money. Hyland, therefore, had not been surprised when they wanted to talk to him.

"Harold got up to the easel, like that, and drew all the stuff out and explained it. ... I think I understood what he was talking about pretty well, although I can't describe it ... and they seemed pretty confident about what they were doing ... and they told me it had to go up on the equator, or very near to it. ..."

Hyland realized that Rosen and Williams were saying they would need to put a launch site on Christmas Island. Recalling that Christmas Island was British territory, he seized on this as his defense, pointing out that Britain was not notable as a place that let people in, especially outsiders that might be in competition with them.

"... you can't get on the damn island with any heavy equipment; it's a rockbound coast, and it's going to be very hard to get at, and it's going to take a lot of money, and furthermore ..."

Hyland was, he remembers, beginning to like the sound of his own voice, "talking about these immense things, you know, and I convinced myself that what I was saying was true, and I thought that I had convinced them." Rosen and Williams retreated.

And regrouped.

Hyland, in the meantime, was not happy at having discouraged two young engineers with innovative ideas. He needn't have worried. Shortly afterwards, Williams turned up with a check for $10,000, saying that this was his contribution to the development of the communications satellite and that colleagues wanted to contribute too.

"Well," said Hyland, "this posed a real problem because this guy was really serious. I had met him two or three times in the interval; he was a great guy, a magnificent mind, and I was kind of flabbergasted inside, but I couldn't do anything about it because I knew this guy was serious and that somehow or other I'd have to put it in a palatable form." Hyland had learned "how to do a hell of a lot of talking without saying anything," and he did so now, stalling until he made a decision. In making that decision Hyland says that he found out what his job in running the company should be.

"I could no longer keep up with them. I was a pretty good engineer in my day and time ... but the art had gone beyond me, and the contribution I could make was to provide ... an environment in which young guys like that should work. I was no longer capable firsthand of making decisions of that sort. They had made the decisions about what they could do, and I had to back it up or deny it. So, I decided to back it up."

And Hyland did back the project. Rosen and Williams were vindicated. Their satellite was built and launched. The Hughes Aircraft Company is now the world's largest manufacturer of satellites.

How accurate is Hyland's recollection of events? It contains a lot of truth, in essence if not literally: Without Hyland's support, the satellite would not have progressed beyond paper studies; Rosen and Williams were brilliant engineers; they (and Tom Hudspeth) did offer to invest $10,000 of their own money in the development; and Hyland did allow engineers room to do their job.

The whole story, though, is both more and less colorful, and Rosen and Williams needed far more persistence than Hyland's recollection demonstrates.

The Hughes Aircraft Company was not one of the first to see commercial promise in the space age. The company's executives watched with amusement, perhaps with a little schadenfreude, the tribulations experienced by the Vanguard team. Even after *Sputnik I* was launched, satellites did not immediately seem to promise great business opportunities. Their launch vehicles failed routinely. Satellites that did reach space did not attain their nominal orbits. They tumbled. Their components failed.

But at the beginning of 1958, the Advanced Projects Research Agency was formed, and later in the year, NASA came into being. Companies like RCA, Lockheed, General Electric, Ford Aerospace, and Philco were exploring the opportunities that satellites offered. Scientists and the Department of Defense were keen to exploit the new frontier. The Soviet Union's achievements challenged the nation's sense of itself.

So by 1959, when Harold Rosen was asked to think of new business ventures to replace the radar contracts that Hughes was losing, space was an obvious area to consider. Rosen discussed the situation with Tom Hudspeth, who as a keen amateur radio operator knew of the parlous state of international communications and of the upcoming solar minimum. Rosen talked, too, with John Mendel. Both wondered whether communications satellites might not be the thing to get involved with. Both men were to make crucial contributions to Rosen's proposal for a twenty-four-hour satellite.

Rosen also talked with Don Williams, whose contribution to the twenty-four-hour satellite was to be the subject of a thirty-year patent battle between Hughes and the government, Intelsat, and Ford Aerospace. Hughes won the battle in the mid 1990s.

Williams, by all accounts, was brilliant, technically very broad, but socially a little narrow. He saddened colleagues and friends when he killed himself in 1966. Bob Roney, who recruited him to Hughes, said, "You asked me if I could remember where I was when *Sputnik* was launched. I can't. But I remember that day, when I heard about Don."

Williams applied to Hughes from Harvard. The position he was interested in had already been filled, but Roney, after reading his résumé, decided that "this was not the sort of person you waited until a vacancy came up to employ." Roney offered Williams a job. But then Williams, whose legacy of memos and technical notes suggest an acutely active and restive intellect, left Hughes to set up his own business with an entrepreneurial friend. There was some story at the time about bugs in Coca Cola bottles, recalled Roney. Williams and his friend developed an inspection device for bottles and were doing quite well.

Rosen watched and waited. When he sensed that Williams was ready to return to Hughes, Rosen set about enticing him back. Hughes had heard from its Washington office that there was a need for radar to detect Soviet satellites. "We wanted to build a giant radar that would look up into the sky and determine an orbit very quickly, and I knew that Don, who was a wonderful mathematician, would be really great at this work. And besides, he was the only one I knew who had any training in astronomy— it was a kind of astronomical problem. So I figured he'd be good for the job, and that's how I lured him back into the company. I told him that space was really hot."

Nothing came of this project, but Williams became interested in navigation satellites and started to think about geostationary orbits. This was a problem that interested Allen Puckett, one of Hughes's senior executives, who would take over the company when Hyland retired.

Rosen, in the meantime, had searched the sparse literature on communications satellites and had found an article by John Pierce that predicted that it would be decades before communications satellites could operate from geostationary orbit. Pierce was perhaps hampered by his more intimate knowledge of the unfolding debacle that would be Advent and his concerns about whether people would find the time delay and echoes intolerable. Unhampered by these doubts, Rosen was convinced that the job could be done much sooner. Knowing of Williams's interest in geostationary orbits he went to talk to him, and the two began working on some ideas for a twenty-four-hour satellite. Thus a formidable engineering partnership was born, with Rosen as the senior partner. Rosen decided that Hughes should develop a small, lightweight, spin-stabilized communications satellite for a twenty-four-hour orbit that could be developed and launched in a year on an existing, comparatively cheap launch vehicle. His suggestion today would be about as revolutionary as observing that cars might have a significant role to play in transportation. In September 1959, his idea was provocative.

The most important decision Rosen made was that the satellite would attain and maintain its stabilities by spinning.

The only other method of stabilizing a satellite in a twenty-four-hour orbit is by spinning wheels arranged internally on each orthogonal axis—three-axis stabilization. For various technical reasons, which hold true for satellites constructed with *today's* technology, three-axis stabilization is the better choice for large satellites in a geostationary orbit. Even Hughes, which built its reputation by taking spinners to their design limits, now makes three-axis stabilized communications satellites.

To the military, which in 1959 was sponsoring the only twenty-four-hour satellite, three-axis stabilization seemed like a good idea. A three-axis stabilized, twenty-four-hour satellite keeps its antennas pointing directly toward the earth and its solar cells oriented towards the sun. Thus high-gain directional antennas can be mounted, and all the solar cells provide power. By contrast, a spinner in 1959, before the days of de-spun antenna platforms, needed an antenna that radiated a signal in all directions, thus dissipating a lot of power to space, and because it was spinning, only about a third of its solar cells could be directed toward the sun at any time.

On the face of it, then, there were good reasons for the Army's decision to make its twenty-four-hour satellite a three-axis stabilized spacecraft. Nevertheless, the choice was a poor one given the technology of the day. First, three-axis stabilized spacecraft below a certain size are more complex than comparable spinners. Second, the twenty-four-hour satellite (which would be called *Advent*) relied on triodes, which are weighter than transistors. Finally, the limitations of the existing launch vehicles and guidance and control made weight an even more critical consideration than it is today.

By choosing spin-stabilization, Rosen automatically saved weight compared with a three-axis stabilized satellite of comparable size. There were immediate weight savings—in the thermal subsystem, for example.

Having decided on a spinner, Rosen was left facing the difficult issue of how to provide a detectable signal from an omnidirectional antenna at geosynchronous altitude. What they realized was that they could provide a usable signal if they selected an antenna that radiated a signal like a giant doughnut rather than the spherical signal of an omnidirectional antenna. Such an antenna would yield higher gain than an omnidirectional antenna even if the gain were not as high as that of the focused antenna that can be mounted on an three-axis stabilized spacecraft.

If the signal were to be usable, however, the satellite, which would be spinning on injection into orbit, had to be stopped (de-spun), turned through ninety degrees, and spun up with its antenna correctly oriented with respect to the earth. Then it had to be moved into position and to keep that position (station keeping). One of the ingenious aspects of the Hughes twenty-four-hour satellite was how this attitude control and station keeping were achieved, and the enabling technology was the subject of the controversial Williams patent.

Williams' idea took advantage of the fact that the satellite was spinning. The Williams patent describes a satellite with two thrusters, one radial and one axial. These could be controlled from the ground and instructed to expel compressed nitrogen, say, during carefully calculated portions of successive revolutions. These spin-phased thrusts would thus move the satellite to the desired attitude or position in orbit. The spin-phased pulses were Rosen's idea, but it was Williams who developed the concept into a feasible technical solution.

In addition to being lighter and simpler than the elaborate system of station keeping and attitude control employed by three-axis stabilized spacecraft, the Rosen–Williams approach had no need of a complex system to deliver the fuel to the thrusters, because the satellite's centrifugal force did the job.

That Rosen should consider a spinner was not that surprising. In his days at Caltech, which he had selected not because of its academic reputation but because of an article in *Life* about Southern California beach parties, Rosen became intrigued by the dynamics of spinning bodies. Sid Metzger, then at RCA, who later headed Comsat's engineering division, said that when RCA's engineers heard about the Hughes spinner they could have kicked themselves for overlooking this approach.

The electronics in the Rosen–Williams proposal were equally important. First, John Mendel suggested that the traveling wave tube's magnet should comprise a row of tiny ceramic magnets, which would weigh less than a solid magnet. Tom Hudspeth's goal was to ensure that this was the only tube in the satellite, which in 1959 was a tall order. His toughest job was finding a way to provide the local oscillator that converted their uplink frequency of five hundred megacycles to a downlink frequency of two kilomegacycles (more familiar today as two gigahertz). Transistors did not work at these frequencies, so he used transistors that operated at lower frequencies and designed a cascade of frequency multipliers into what Rosen calls "a genius geometry." Hudspeth is a reticent man who says little about those early days and even finds it depressing to talk of the past. Nevertheless, he still had one of these early frequency multipliers in a brown paper bag under a desk in his lab.

Rosen and Williams wrote up their proposal for a twenty-four-hour satellite in September 1959, the same month that Leroy Tillotson sent his proposal for a medium-altitude active repeater to Bell's research department. On September 25, Rosen's immediate superior Frank Carver, who had asked Rosen to think of new business ventures, took the proposal to Allen Puckett, a senior executive. Though not immediately convinced, Puckett did not kill the idea. In October, Rosen and Williams's proposal, "Preliminary design analysis for a commercial communication satellite," was circulating internally. It contained the principles that would become Syncom, though they were embodied in what would seem to today's eyes to be an unfamiliar design. The satellite, a spinner, was to be a cube seventeen inches on each side because, because, they argued, a cube was easier than a cylinder to construct. It would be equipped with a gun and bullets or powder charges capable of imparting four different thrusts for station keeping. The gun would be triggered from the ground at the moment in a revolution that would impart the necessary velocity correction. The gun could be either "an automatic type firing multiple shots from a single barrel, or a multiple-barrel device using electric primers." An amended proposal envisaged bullets or charges capable of imparting sixteen increments of thrust. Not until early 1960 did the satellite begin to look familiar to a modern eye. By then it was cylindrical and expelled compressed nitrogen for attitude control and station keeping.

They were not sure in their first proposal whether they needed to correct for lunisolar gravity, but they had calculated how much the satellite would drift if they did not get quite the right velocity for a geosynchronous altitude.

Thomas Hudspeth, Harold A. Rosen and Donald D. Williams pose with the *Syncom* satellite they pioneered and which led to the era of commercial communication satellites. Dr. Williams holds the travelling wavetube that was a crucial component of the satellite.

The satellite, they said, would weigh twenty pounds, be developed in a year, and cost $5 million. It would have sufficient bandwidth for either TV transmission or 100 two-way telephone channels. The weight grew during the next few years, but still *Syncom* was about a tenth of the projected weight of *Advent*.

Like Pierce, Rosen wanted to exclude the government and to keep the twenty-four-hour satellite as a private business venture. The proposal suggested that Hughes should build a launch site on Jarvis Island (not Christmas Island), close to the equator, and buy some Scout rockets to launch the satellite.

"When Harold came up with the idea," said Roney, "there was internal tension. Some people in the communications lab thought it should be theirs, not over in the radar lab." But not Samuel Lutz, who headed the communications lab. "He was," says Roney, "fascinated by Harold's work." He was also deeply ambivalent.

During 1956 and 1957, Lutz had presented his own ideas for communications satellites to Hughes's executives. Amused as they were by Vanguard, the executives concluded that Lutz's ideas were romantic. He was, however, an obvious person to examine the concepts outlined in the Rosen–Williams proposal. From the beginning he saw the innovativeness of what Rosen and Williams were doing, but then he would fret that the design was "far from conservative." Sometimes he saw market opportunities; at other times he was skeptical of Hughes's involvement in a field that AT&T dominated with such assurance. As for Rosen's plans for live international television, he had profound doubts.

"Undeniably," he wrote on October 1, 1959, "they [live TV programs] would have novelty and propaganda value, and there always would be occasional events of international interest. Many race-crazy Europeans would stay up all night to watch our Indianapolis races, while some of our wives might get up before breakfast to watch a live coronation or royal wedding—but not very often. ... Most of the few programs of international interest are already being flown by jet or transmitted at slowed down rates via cable, with the time difference in its favor. Thus Rosen's estimate of an hour per day of TV revenue appears optimistic. An hour per week seems more realistic. ..."

Nine days later, Lutz was one of five people given two weeks to review the proposal. Rosen was the chairman of the small panel. By October 12, they had made good technical progress but were struggling with the economics. On October 22, they concluded unanimously that the project was technically feasible in close to the time and price suggested and that it should not encounter any legal or technical problems. Mendel's assurance that the travelling wave tube was feasible won Lutz's backing.

They said that the economics needed further study, but that population increase, the shrinkage of travel time via commercial jet, increasing foreign industrialization and international commerce, and the forthcoming decrease in high-frequency communication because of solar minimum all made the proposal economically attractive.

Four days later the plan was put to Hyland. He ordered "an immediate and comprehensive study of patentability" and an inquiry to determine NASA's position with respect to commercial rights. From the beginning Hyland wanted the satellite to be a government rather than private project.

On October 29, 1959, Hyland learned that there might be patentability in the attitude control and station keeping method, and the company's lawyer advised that it should be reduced to practice before any presentation was made to NASA; otherwise, NASA might seek to patent anything made during a contract with them.

On November 2, Rosen and Williams signed an invention disclosure that stated, "a series of discrete impulses obtained from the recoil of a multiple shotgun are used to provide vernier velocity control and position adjustments." Three days later, Williams traveled to Washington to brief Homer Joe Stewart. Williams, wrote Edgar Morse in a NASA historical document in 1964, was "very conscious of Hughes commercial plans and began by establishing that Hughes would not lose its proprietary and patent rights by having the discussions." Williams's caution, as three decades of litigation show, was well placed.

On his return to Culver City, Williams immersed himself in work stemming from Stewart's critique. Allen Puckett approached Roney to conduct another review. Puckett was fascinated but was hearing technical criticisms and cautions that communications satellites were a bad idea politically for Hughes. Could he, Puckett asked Roney, stake his reputation on this idea?

To men of Rosen's and Williams's disposition these and other studies must have been a sore trial. Rosen was not a diplomat. Even though the satellite was his idea, it was Allen Puckett, not Rosen, who carried news to congressional hearings. "No one in their right minds," said Roney, "would have let Harold testify. He was volatile." And once the twenty-four-hour satellite had become Project Syncom, C. Gordon Murphy was the project leader, Rosen the project scientist. Gordon Murphy was needed, it seems, because by then, "Harold had alienated so many people in Washington." The memory, when Roney talks, is clearly affectionate.

Nor was Rosen much more conciliatory within Hughes. When the company decided in January 1961 that it would respond to NASA's request for proposals for *Relay*, Rosen would have nothing to do with it, nor with the company's bid for *Advent*. Williams was equally unbiddable and bombarded Hyland with memos critical of the company's policy.

As 1959 turned to 1960 what patience they had was already being severely tested. They had up until then done their work with discretionary funds. To go further they needed the company to adopt the project.

On December 1, 1959, Hyland decided not to commit funds "at this time." Rosen, Williams, and Hudspeth were not acquiescent, nor were they clear about what he could do, but they decided to put up $10,000 each of their own money and to seek outside support. "I invited John Mendel to join us but he didn't have the, uh, he said he wasn't gutsy," said Rosen. They tapped any contacts they could think of. Rosen had a friend who was the chief engineer at Mattel. "Mattel had just come in to a lot of money with a Barbie doll. And I thought they might be looking for some good investments. It turned out they weren't. They invested in Ken instead." Hudspeth had a more likely contact in the aerospace industry who had hit it big on some company. "That was frustrating, because he led us on a little bit, but he was really full of hot air," said Rosen.

No one was biting. "Those were the days," recalled Rosen, "when our boosters used to blow up in front of the television regularly." Rosen "stewed and brooded." Then he contacted his old boss, Tom Phillips, at Raytheon. In February 1960, Rosen and Williams flew to Boston. Yvonne Getting, later head of the Aerospace Corporation, was there, and he was skeptical about the whole idea. "He didn't even want us to talk to him about it, because he thought we'd get involved in future patent disputes. He eventually listened, but I don't think he was very enthusiastic."

But Tom Phillips was, and he told Rosen and Williams that if they would come and work for him, he would give the project his personal attention. Phillips was, at that time, on his way to becoming the president of Raytheon. They also met Charles Francis Adams, who was running Raytheon. Phillips tweaked Rosen and Williams about Howard Hughes, saying, "here you are talking to the president of Raytheon when you haven't even seen the president of your own company."

Back in Culver City, Rosen told Frank Carver that he was going to work for Tom Phillips. Carver immediately took him to see Allen Puckett, who took him to see Pat Hyland. Who knows whether Rosen's machinations won the day, but, says Rosen, "Mr. Hyland said he was going to support us right here at Hughes, which was great. I was really happy to hear that." When Rosen told Williams of Hyland's response, Williams sent his cheque for $10,000 to Hyland asking whether he could invest his money in the new Hughes project.

Hyland authorized an in-house project to develop the spacecraft and the traveling wave tube amplifier on March 1, 1960. He would not order a booster or sign a contract for a launcher. All the same, Williams took a trip to Jarvis Island, where "he got some very nice photographs of birds," recalls Rosen.

When Hyland authorized the in-house project, Rosen, Williams, and Hudspeth asked the company to release them from their usual patent agreement if Hughes decided not to develop the satellite. Puckett was sympathetic. But in the end, there was no need to release them. Once they were committed to the project, Hyland and Puckett were wholehearted in their support and in their efforts to sell the satellite to the government.

20 Syncom

"We had from our first presentation improved the control system quite a lot and we continued to improve it over the years. We got it down to two thrusters; then we started to work on the precession capabilities of one thruster. I had at first thought it would take four, but Don had come back very quickly with a major improvement. We had four for redundancy, but we still could precess it with one jet. NASA was skeptical about the single-pulse thruster." Thus Rosen thirty years later.

NASA, which had some contact with Hughes in February 1960, was skeptical about considerably more than the single-pulse thruster. Rosen had proposed that the inexpensive Scout rocket, which Richard Kershner also favored for Transit, should launch the twenty-four-hour satellite. In the same month that Hyland authorized an in-house program, NASA's Langley Research Center completed an analysis of the Scout rocket, and concluded that it could not place a twenty-five-pound satellite (the Rosen–Williams satellite had now grown from twenty pounds and would put on more weight, but not as much as *Advent*) in a twenty-four-hour orbit. Nor could any other launch vehicle.

But Puckett and Hyland were now supportive. Puckett coordinated the marketing, keeping Hyland informed of his progress. There were approaches to GTE (a common carrier) and CBS, attempting to promote interest in the idea of a communications satellite, and to E.G. Witting, the director of research and development in the Department of the Army.

On April 1, 1960, Puckett briefed Herb York, formerly the head of the Advanced Projects Research Agency, now the director of defense research and engineering, and John Rubel's boss.

In May, the month that the first attempt to launch *Echo* failed, Williams was immersed in dynamic analyses of the satellite. Rosen was preparing another proposal, specifying Thor-Delta, which had a greater lift capability than Scout, as the launch vehicle. Hughes sent an unsolicited copy of the revised proposal to NASA on June 15, 1960. NASA, however, was confined to work on passive satellites at the time by its agreement with the Department of Defense.

In July, AT&T disclosed its $170 million plan for a constellation of fifty medium-altitude satellites and argued that the Federal Communications Commission should reserve frequencies for satellite communication. Williams was as usual preoccupied with engineering, this time concentrating on calculations of moments of inertia. Hughes made a presentation soliciting support to the space science panel of the president's science advisory committee. They received a polite letter of thanks, but no encouragement.

In the meantime, Hughes' man in Washington had also been busy. He knew one of Vice President Nixon's bodyguards, and the vice president, said Rosen, owed the bodyguard a favor. He arranged for Hughes to make a presentation to the chairman of the Republican National Committee, who told them that they looked like nice people—surely they didn't want to get mixed up with politics. They did, however, get a meeting with NASA administrator T. Keith Glennan out of the encounter. "So you can see," said Rosen, "we were grasping at straws."

In August, the policy confining NASA to work on passive communications satellites was formally abandoned. Within days, John Pierce and senior AT&T executives briefed NASA on Bell's experimental active satellite plans. On the same day, Hughes executives, by now contemplating a joint venture, had another meeting with GTE.

A few days later Puckett, in the meeting that resulted from the encounter with Hall of the Republican National Committee, briefed the NASA administrator T. Keith Glennan. Glennan said that Puckett was talking through his hat and recommended that Hughes work on medium-altitude satellites.

Glennan's comment must be judged in context. NASA's Langley Research Center concluded that the Hughes proposal was marginal, and the new agreement with the Defense Department, signed only days earlier, would soon preclude NASA from involvement in communications satellites in twenty-four-hour orbits.

Despite Glennan's negativity, NASA's head of communication satellites, Leonard Jaffe visited Hughes on September 1. To Samuel Lutz, it looked as though Rosen and the supporters of a twenty-four-hour satellite now thought that they had a NASA contract in the bag. If they did, they either did not know some important things, for example the agreement between NASA and the Defense Department, or there was a high degree of wishful thinking.

Meanwhile, Puckett was hearing negative views from colleagues within the company. Critical memos reached him saying that Hughes was presenting a fragmentary approach to potential military customers and that the best role for Hughes was as a supplier to AT&T. On the other side of the country, at AT&T's corporate headquarters in New York, the issue of the Hughes repeater surfaced at a meeting of Bell's engineers on September 1, 1960. A report of the meeting says "Hughes is supposed to have an ingenious twenty-pound repeater (TV bandwidth is claimed)." Those at the meeting agreed that if the repeater was any good, Bell should try to obtain it for trial.

The supporters of the twenty-four-hour satellite within Hughes continued to try to involve GTE, and GTE's technical staff spent four days at Hughes in early September. They made no commitments, but Rosen sensed that GTE would join Hughes. At Hyland's request, Hughes briefed the RAND Corporation. Puckett warned his people to say nothing about the discussions with GTE. And later that September, John Rubel, who had worked at Hughes before moving to the Office of the Director of Research and Engineering in May 1959, visited his old workplace. Puckett had already briefed Herb York about the Hughes twenty-four-hour satellite in April, and presumably, Rubel now learned more.

On October 2, in a meeting that must have been full of tension, Rosen, Williams, and Hudspeth briefed the engineers at Bell. Among those present were John Pierce and Leroy Tillotson. As Pierce listened he became convinced that Rosen was a wild-eyed dreamer, willing to say anything to sell his satellite.

Shortly afterwards, Witting, the Army's R&D director, wrote to Hughes. He was unapologetically dismissive of the Hughes proposal and fluent in his condemnation of the weaknesses of many of the satellite's systems. Rosen was infuriated. He called Witting and wrote a deeply sarcastic letter, pointing out the errors in Witting's evaluation. Just under a year later, Witting's successor would respond to this letter, reaffirming everything that Witting had said. It was written on the day that NASA, supported by the Department of Defense, placed a sole-source contract with Hughes for a twenty-four-hour satellite.

In November 1960, Puckett, Rosen, and Williams briefed ITT, General Bernard Schriever, the British military, and Stanford Research Laboratory. At the Cosmos Club in Washington D.C., Puckett talked enthusiastically of the proposal to Lee DuBridge, the president of Caltech. But, despite the flurry of technical and marketing activities; the optimism and confidence of Rosen, Williams, and Hudspeth; and the support of Hyland and Puckett, the Hughes Aircraft Company was by the end of the year no further forward. No one wanted their satellite. GTE had made no more moves towards a joint venture. By the next spring, GTE would be in partnership with RCA and Lockheed, and that group would be proposing its own ideas for satellites in a twenty-four-hour orbit. Hughes had no contract, nor any likelihood of a contract, with NASA, which was bound by its agreement with the Defense Department not to develop twenty-four-hour satellites. The Army had rejected them, and in such a comprehensively dismissive way that there could be no hope remaining from that quarter. The Air Force said that the proposal was marginal and overoptimistic with regard to payload capability. These two branches of the military were engaged in one of their frequent skirmishes, with control of the development of communications satellites the disputed territory. Rosen's proposal would not have been welcome.

Rather desperately, or so it seemed to some within the company, Hughes announced late in 1960 that it had an off-the-shelf satellite for sale. During the three months at the beginning of 1960 when Hyland had had the project on ice, Rosen, Williams, and Hudspeth had continued their work. Hudspeth had worked on breadboards for the electronics, while Rosen and Williams refined the mechanical structure and ideas about the satellite's control mechanism. As soon as Hyland had given the go-ahead, they'd put together a little project lab and started making things. In May, Hughes had begun construction, and by the fall, they'd demonstrated the control mechanism in the lab and had tested the satellite's ability to transmit television signals. Thus, with their ideas embodied in hardware, they sought a wider audience. They demonstrated the satellite in December at a meeting of the American Rocket Society.

During the winter of 1960–61, very few people were working on the satellite. Bob Roney wrote to Puckett seeking assurances that there would be money in the coming year for further development. Puckett, however, was under pressure to switch the Hughes effort to medium-altitude satellites and to bid in the forthcoming NASA competition for Project Relay. NASA put out its request for proposals in January, and with the same reluctance that characterized AT&T's response, Hughes prepared to compete. Rosen and Williams remained aloof.

Yet the tide of their fortunes was changing, had probably been changing all through that dismal Christmas in a way that is discernable only with hindsight. Between October 1960 and March 1961, three Centaurs blew up, which set back the Centaur development schedule and thus the schedule for *Advent,* the Army's twenty-four-hour satellite. There were several downward revisions of the amount of payload that Centaur could lift, yet *Advent* was getting heavier. John Rubel, of DDR&D, was unhappy about these things and knew that the Department of Defense needed improved communications. Rubel had also visited Hughes in September 1960, and must have learned more about the twenty-four-hour satellite, and he respected Harold Rosen's work. Together these events and Rubel's attitude must have influenced circumstances in favor of the Hughes satellite.

An early indication of the changing tide came when Rosen was asked to brief the Institute for Defense Analysis (IDA) on January 11, 1961, about the Hughes twenty-four-hour satellite. The IDA was evaluating communications satellites for the Office of the Director of Research Engineering. Rosen's report of the meeting, written the next day, records that the IDA made generally favorable remarks about his presentation and was critical of *Advent.* He wrote that one panel member had said that program managers apparently placed more faith in the development of the Centaur rocket than of a traveling wave tube, and that the panel member did not consider the attitude justified.

This comment alluded to the reservations some felt about John Mendel's prospects of successfully developing his lightweight traveling wave tube. In February, Rosen, in response to questions for Rubel, sent a telegram saying that the traveling wave tube had been chosen because of its superior performance. He cited publications by Bell and comments from John Pierce to bolster his case. If there were to be problems with the tube, Rosen wrote, it could be replaced with a triode even though the satellite would then operate at reduced power and bandwidth.

In California, Williams was preoccupied by the recently discovered tri-axiality of the Earth and its influence on the motion of satellites in geosynchronous orbit. During that same February, Hughes executives were discussing what the company should do if it did not win the Project Relay competition. And Samuel Lutz, at Puckett's request, was again reviewing the twenty-four-hour satellite. This time Lutz was much more negative. The satellite, Lutz wrote, had not shown "the high degree of engineering conservatism which would give it sales appeal to the common carrier." Competitors, he pointed out, offered satellites with a longer life for very little extra delay in development. Lutz recommended that no effort be spared to win the NASA competition for Relay. If successful, the company would save face and recover some of its half-million investment. Lutz's report clearly shows that he was intimidated by AT&T's monopoly position, by its financial resources, and by the technical resources of Bell. "Do we," he asked, "want a future in this field badly enough to make the effort it will require?"

Even as the words clattered out of his typewriter, the moves were being made that would set the Hughes Aircraft Company on its path to *Syncom, Early Bird* and a preeminent position among satellite manufacturers. The hard work of balancing out on a limb had, though they did not know it, been done.

At the end of March 1961, Hughes made a presentation to Rubel in Washington. For the next few months Puckett and other Hughes executives would hang on Rubel's every word. If they talked with him over dinner or at a meeting, a memo was circulated, reporting either what was said or what they read between the lines.

Puckett had learned from Rubel that the administration was concerned that the country was not moving quickly enough toward a communications satellite capable of either military or commercial operation. The current plans, Rubel had told Puckett, were for a system that was "a long way downstream" as well as "very expensive," and it would be appropriate to seek an interim system. Puckett had asked what Hughes could do, and Rubel had replied that the Department of Defense had received proposals in varying degrees of formality, but had no reasonable means of choosing among them. Hughes, he said, could perhaps produce a white paper providing a historical and technical context for a decision.

On May 8, Puckett sent Rubel a letter, making no reference to their previous discussion. In it Puckett wrote that Hughes had prepared a special research study dealing with various aspects of the military communication problem. The study's purpose, he wrote, was to examine the possible value of a lightweight spacecraft as an interim communications satellite. Puckett offered to submit a full proposal "if you believe this deserves continued consideration." Hughes was at this stage close to having completed the proposal mentioned, and Puckett already knew from C. Gordon Murphy what would be an acceptable date for submission of the proposal to DDR&D.

It seems that the contents of these discussions did not filter down to Williams, who on May 11 wrote a long, critical memo to Hyland. It began, "You are aware of my bullish outlook regarding commercial communication satellites and my confidence in the Hughes stationary satellite concept. For the past several months, I have been concerned that Hughes is letting its technical advantage slip away for political reasons...." He wrote persuasively and at length, but his views were at odds with company policy because he still hoped that Hughes would undertake the project without government involvement.

In Washington events moved apace. The policy debate provoked by monopoly considerations and by Berkner's assertion that communication satellites would be a billion dollar business was well underway, and a decision had been taken that *Advent* would continue. But Hughes's star was still rising.

On June 6, 1961, events had reached a stage allowing Jaffe to recommend that NASA negotiate with Hughes to develop a twenty-four-hour satellite. Jaffe thought there was no doubt as to the ultimate desirability of twenty-four-hour satellites even though he remained convinced they would not be operational for years.

Given the division of labor that NASA and the Defense Department had agreed to, this recommendation was possible only because Rubel, along with Robert Seamans, NASA's associate administrator, was plotting the agreement's abolition. Seamans viewed the idea of an operational communications system based on tens of medium-altitude satellites as impractical. Where, he had asked Bell, do you propose to get all your computers from? Rubel, of course, knew that the Defense Department needed an interim satellite to provide some cover during the solar minimum, but he judged that the time was not right for canceling *Advent*. If, however, the agreement between NASA and Defense could be set aside, NASA could place a contract with Hughes to explore the alternative technology of a lightweight twenty-four-hour satellite.

On June 17, Rubel was holding discussions with NASA. Another senior Hughes executive, A.S. Jerrems, happened to be in Washington that weekend, and he met Rubel in the evening. Jerrems wrote to Puckett, "He was inscrutable about the detailed content of the meeting, but he made a statement to the effect that, in his opinion, HAC's [Hughes Aircraft Company] proposal for getting a geosynchronous satellite funded are better now than they have ever been."

On June 21 the odds in favor of Hughes improved again. The Institute for Defense Analysis met to discuss the merits of an experiment with a lightweight satellite. The IDA concluded that if the country decided to have only one program in active satellites in addition to Advent, then that program

Thomas Hudspeth (left) and Dr. Harold A. Rosen stand atop the Eiffel Tower during the Paris Air Show of 1962. Between them is the prototype *Syncom* satellite which they and Dr. Donald D. Williams fought so hard for.

should be for medium-altitude satellites. Further, they said that an experiment with lightweight satellites should not interfere with Advent (the Army) or Project Westford (né Needles—the Air Force), nor should it affect the determination to pursue medium-altitude satellites. Having saved everyone's face, the panel said that the experiment with a lightweight active repeater was

Harold A. Rosen (right foreground) and Thomas Hudspeth hold the prototype of *Syncom,* the world's first synchronous orbit satellite. Behind them is *Intelsat VI,* a later generation of communication satellite. The tiny Syncom would fit in one of the fuel tanks which Dr. Rosen is pointing toward.

unique and should be undertaken. Two days later, the deputy secretary of defense, Roswell Gilpatrick, wrote to James Webb effectively releasing NASA from its tacit agreement not to work on active satellites in geostationary orbit.

There was still much for the administrators to do, but the Hughes twenty-four-hour satellite was now secure. On July 27, Abe Silverstein,

NASA's director of spaceflight programs, told Goddard to put together a preliminary project plan for Hughes that was to be prepared with the Army's Advent Management Agency. On August 11, NASA announced that the Hughes Aircraft Company had been chosen on a sole-source basis to build a twenty-four-hour satellite. Goddard decided that the satellite should be called *Syncom* (for synchronous communication). The first launch attempt failed. Somehow, it seems almost obligatory that it should have done so. It was a black day for Harold Rosen—elation followed by despair. The second attempt, on July 26, 1963, succeeded. It was launched into a "quasi-geostationary" orbit, which was easier to reach than a true geostationary orbit: it was at geosynchronous altitude but was not coplanar with the equator. Still, *Syncom* proved that communication was possible via a radio relay at geosynchronous altitude. It had just one voice channel. But together with *Syncoms II* and *III,* it demonstrated the technology and led to the selection of one of Harold's spinners as *Early Bird,* which opened the still unfolding era of global telecommunications.

But all of you have lived through the last four years and have seen the significance of space and the adventure of space, and no one can predict with certainty what the ultimate meaning will be. . . .

—John F. Kennedy, May 25, 1961

Epilogue

We are riding through the outskirts of a jungle in French Guiana. At the front of the bus a woman is instructing us in the use of a gas mask. The mask looks remarkably like a leftover from the First World War, and it seems to me unlikely that any of us will succeed in donning the apparatus should the rocket we are to watch spew noxious fumes in our direction. We are, after all, journalists and will have imbibed several glasses of something interesting by then.

The launch is to be at night, and we are to watch from Le Toucan, an open-air bar in the jungle some miles from the rocket. It promises to be spectacular—a very pleasant junket.

At this time I have never heard of Sergei Korolev, survivor of Kolyma and chief designer of cosmic rocket systems. I do not know of his triumphs and despair nor of his struggle to launch *Sputnik*. I have never heard of the team of engineers who stood with tears in their eyes in a smoke-filled room in Maryland while a satellite transmission faded. I know nothing of Verner Suomi and Robert Parent trapped in a bunker at Cape Canaveral while a rocket smoldered outside. The names John Pierce and Harold Rosen mean nothing to me.

These men and the things they did belong to a time nearly forty years before the launch I am waiting to watch, long before this launch site even existed. It was a time in rocketry when failure was more common than success. Vaguely we journalists know that space is still risky, but we expect in a few hours time to drink a toast to a successful launch. When we do, it will be because of those men and hundreds of others.

The satellite in the nose cone might be American, or perhaps French; maybe it's Saudi Arabian or Indian. It might be a weather satellite or a science satellite or a communications satellite—one of the Hughes Aircraft Company's Galaxy class, which barely clears the doors of a jumbo jet when it is loaded for its flight to South America.

Inside the Jupiter control room, there will be the usual concentrated prelaunch tension. But there will be no slide rules and no teletype machines.

After this launch, no one will inform the Kremlin. No one will call an American president.

The countdown proceeds.

With an inner frisson belying our outward nonchalance, we journalists hear a voice: "dix, neuf . . ."

No lone bugler has heralded this voice, which does not falter as did the voice during the launch of the first *Transit*. Goldstone is not waiting. William Pickering does not have a line open to the Cape. No car waits to whisk anyone through rain-soaked midnight streets to a room packed with the world's press.

But there is silence. And our eyes are fixed on the rocket. We hold our breath. We dare not blink. And then, it happens. Incandescent flames billow around the distant rocket. Impossibly, it struggles upwards, gathers speed, and, as a thunderous roar washes past our ears, the rocket passes to become a distant moon.

Chronology

1945 October. Publication of Arthur C. Clarke's paper on extraterrestrial relays, outlining his ideas for a communications satellite in geostationary orbit.

1946 April 16. The first V2 rocket fired on American soil.

1947 October 18. The launch of a reconstructed V2 from Kapustin Yar. The work is undertaken under the direction of Sergei Korolev.

1950 James Van Allen hosts a dinner party and Lloyd Berkner proposes an international effort to study the geophysics of the earth and its environment. This becomes the International Geophysical Year of 1957/58.

June 24. North Korea invades South Korea

1951 Publication of the RAND report: "Inquiry into the feasibility of weather reconnaissance from a satellite," by William Kellogg and Stanley Greenfield.

1953 March 26. The first meeting of the U.S. National Committee of the IGY. The group defines a program of research and a budget. Satellites are not included.

Fall. John Pierce presents a paper to the Princeton, New Jersey, branch of the Institute of Radio Engineers, outlining three possible types of communications satellite.

1954 July. President Eisenhower appoints James Killian to head a panel of scientists charged with identifying ways to prevent surprise attacks on the U.S.

October. A meeting in Rome of the international scientific community planning the IGY endorses the inclusion of satellites in the IGY.

1955 January. The U.S. National Committee of the IGY asks three experts to assess the feasibility of developing a "long playing" rocket, that is, a rocket capable of launching a satellite.

February. The Killian committee formally reports. One of its recommendations is that the United States develop a scientific satellite to establish the idea of freedom of space.

July 19. President Eisenhower suggests an open skies policy in Geneva.

July 29. President Eisenhower announces that the U.S. will launch a satellite as part of the International Geophysical Year.

August 2. Academician Leonid Sedov holds an informal press conference during a meeting of the International Astronautical Federation in Copenhagen and announces that the USSR will launch a satellite as part of the IGY.

September 9. The secretaries of the Army, Air Force, and Navy are notified that the Navy has been selected to develop Vanguard as a launch vehicle for the IGY.

September 24. President Eisenhower has a heart attack.

1956 January. The Technical Panel on Earth Satellites, chaired by James Van Allen, assesses formal proposals for satellite experiments.

January 30. The USSR Academy of Sciences decides that the R7 rocket will be used to launch a satellite.

March. Hearings begin on Capitol Hill on a supplemental budget for the IGY, including funds for the newly adopted satellite program.

June 7. Harry Wexler presents Verner Suomi's proposal for a radiation budget experiment.

July. Egypt nationalizes the Suez Canal.

October 2. Student demonstrations in Hungary.

October 29. Israel, backed by Britain and France, invades the Sinai Peninsula.

October 31. The U.S. and the USSR vote together against the Anglo-French action.

November 4. Soviet troops reinvade Hungary.

December 3. Wexler and Suomi get the go-ahead for the radiation budget experiment

1957 January. Training sessions begin for the operators of the Minitrack stations and continue through much of the spring.

Verner Suomi and Bill Stroud are now directly competing to have one or the other of their meteorological payloads chosen for one of the four satellites that are expected to be launched as part of the IGY.

February 7. The tenth meeting of the Technical Panel on Earth Satellites listens to a proposal for a reflective, inflatable sphere from William O'Sullivan, from the Langley Research Center.

July 1. The official start of the IGY.

Nikita Khrushchev faces down an attempt to depose him as first secretary of the Communist Party.

August 3. The USSR launches the world's first ICBM—the R7.

August 26. The USSR announces the successful launch of its ICBM.

September 7. The anniversary of Friedreich Tsiolkovsky's birthday. Lieutenant Anatoy Blagonravov and Sergei Korolev attend celebrations.

September 24. President Eisenhower sends federal troops to Little Rock, Arkansas.

September 30. The opening of an IGY conference on rockets and satellites in Washington D.C.

October 2. Sputnik is moved to the launch pad. **The USSR turns down Japan's request that it stop nuclear testing, saying the U.S. and U.K. must also stop.**

October 3. Sputnik countdown begins.

Friday, October 4—Sputnik is launched.

Monday, October 7. Bill Guier and George Weiffenbach, from the Applied Physics Laboratory of the Johns Hopkins University, tune into Sputnik's signal, "for fun." Within weeks they are developing a new approach to orbital determination.

October 10. The National Security Council meets to discuss the Soviet satellite.

November 3. The launch of Sputnik II.

November 7. President Eisenhower appoints James Killian as special advisor to the president for science and technology.

November 8. The defense secretary, Neil McElroy, directs the Army to prepare for a satellite launch as part of the IGY.

November 8. The director of the Applied Physics Laboratory endorses Guier and Weiffenbach's work, allocating them $20,000.

December 6. Vanguard explodes.

1958 January 9. President Eisenhower makes his first post-Sputnik State-of-the-Union address.

Friday, January 31—Explorer I is launched

February. The Advanced Projects Research Agency formally opens its doors.

February 6. The Senate appoints its first special committee on astronautics, headed by Lyndon B. Johnson.

March 5. President Eisenhower approves a memo directing the Bureau of the Budget (the predecessor of the Office of Management and Budget) to draft a space bill before the Easter recess.

March 17. The launch of *Vanguard I*, the satellite which generated the data showing Earth to be pear-shaped.

Frank McClure tells Guier and Weiffenbach of his ideas for a satellite navigation system.

Spring 1958. John Pierce reads an article about O'Sullivan's plans for launching an inflatable reflective sphere and realizes that it could serve as a passive relay for communication signals.

May 26. An IGY working group shelves O'Sullivan's plans.

July. John Pierce and Rudi Kompfner attend a meeting at Woods Hole, Massachusetts to discuss communications satellites. Pierce talks to William Pickering, of the Jet Propulsion Laboratory, about the possibility of using O'Sullivan's balloon as a passive communication relay, and Pickering raises the idea at the meeting.

August 19. T. Keith Glennan and Hugh Dryden are sworn in as administrator and deputy administrator of the National Aeronautics and Space Administration.

October 1. The National Aeronautics and Space Administration officially opens its doors.

October. John Pierce delivers a paper entitled "Transoceanic Communication via Passive Satellites" to a national symposium on extended range and space communications. The Advanced Projects Research Agency holds a one-day meeting on twenty-four-hour communications satellites (better known today as geostationary satellites). NASA enters an informal (but binding) agreement with the Department of Defense that the agency will develop passive communications satellites while the Pentagon develops active satellites.

December 31. NASA takes over Project Vanguard.

1959 February 17. Launch of Bill Stroud's experiment to determine cloud cover. The satellite goes into orbit, but precesses out of control because of a problem with the launch vehicle.

March. John Pierce and Rudi Kompfner write an internal memo at Bell Laboratories on the feasibility of broadband satellite communications via active repeaters.

August 24. Leroy Tillotson at Bell outlines plans for a medium-altitude active satellite in an internal memo. Much of what he describes becomes Telstar.

September 17. The launch of *Transit 1A*. It fails to reach orbit.

October 13. The launch of Verner Suomi's radiation balance experiment.

October 25. At the Hughes Aircraft Company, Frank Carver presents a proposal by Harold Rosen and Don Williams for a lightweight twenty-four-hour satellite to Allen Puckett, a senior Hughes executive.

December 1. Pat Hyland, chairman of Hughes, decides not to "commit funds at this time" to the development of the Rosen–Williams proposal.

1960 January. Bell scientists and engineers conclude that communications satellites will have to be active. They are interested in a constellation at medium altitudes.

February. Harold Rosen, Don Williams, and Tom Hudspeth decide to put up $10,000 each of their own money to develop their twenty-four-hour satellite idea. They seek, but fail to secure, alternative sources of funding.

March 1. Pat Hyland reverses his decision and approves an internal company project to develop a communications satellite and a new traveling wave tube.

April 1. The TIROS weather satellite is launched.

April 13. The successful launch of *Transit 1B*. The Transit team begin to see how complex the earth's gravitational field is.

April. T. Keith Glennan seeks help from John Rubel, deputy director of defense research and engineering, to reverse the communications satellite policy agreed to by NASA and Defense in October 1958. NASA wants to develop active communications satellites.

Spring and Summer 1960. Bell shifts its efforts to medium-altitude active satellites. In the meantime, NASA (constrained by its agreement with the Department of Defense and dubious about the technology) does not support the Hughes Proposal. The Army and the Air Force are highly critical. No other industry will join Hughes to develop its satellite. Hughes supports its communications satellite work internally.

August. The Department of Defense agrees that NASA can develop active satellites (thus changing the agreement of October 1958). The unwritten proviso understood by both parties is that NASA will develop medium-altitude satellites, while Defense works on twenty-four-hour satellites. The Defense Department's twenty-four-hour satellite is called Advent.

May 13. First Echo launch attempt fails.

August 12. Echo launched successfully.

October. T. Keith Glennan precipitates a policy debate about communications satellites and says that the agency will be sponsoring development of medium-altitude communications satellites.

December. Hughes still has no external support for its satellite ideas. Several people within the company are getting cold feet and urging, for both policy and technical reasons, that Hughes not develop communications satellites.

1961 January. John F. Kennedy is inaugurated as president.

January. NASA puts out a request for proposals for a medium-altitude satellite—Project Relay.

May. RCA wins the contract for Project Relay. Hughes and Bell were also competitors.

James Webb notifies Fred Kappel, of AT&T, that the agency is prepared to launch the company's satellite.

Spring. John Rubel, concerned by development problems with *Advent* and its launch vehicle, the Atlas-Centaur, plans with Robert Seaman at NASA to have the agreement confining NASA to medium-altitude active satellites set aside so that the agency can place a contract for the Hughes satellite.

By summer, the agreement between NASA and Defense has been set aside, and the agency has completed difficult negotiations with AT&T.

By late summer, NASA is supporting Project Relay, has agreed to launch *Telstar,* and has a contract with Hughes for a twenty-four-hour satellite.

All summer a fierce policy debate rages about international communications satellite policy.

1962 February. The White House sends the Communication Satellite Bill to Congress, establishing a private company under close governmental regulation, with fifty percent of its stock allocated to the general public and fifty percent to common carriers.

July 10. Telstar is launched.

August 31. President Kennedy signs the Comsat Act.

December 13. Relay is launched.

1963 February 14. *Syncom I* is launched, but communication is lost on orbital injection.

May 7. *Telstar 2* is launched.

July 26. *Syncom II* is launched successfully into a "quasi-geostationary" orbit.

November 22. President Kennedy assassinated.

1964 July 1964. The formation of Intelsat.

1965 April 6. *Early Bird* is launched.

Notes and Sources

The relative importance of written and oral records varies from section to section and even within sections, as does the balance between primary and secondary sources. In chapter one, for example, secondary sources were the only ones I had access to.

Even when there are records, they can be scanty or one-sided. The IGY, for example, is well recorded by the National Academy of Sciences, but the individual scientists, such as Verner Suomi, do not have extensive records. Often, the scientists and technologists were too busy as pioneers to record in detail what they were doing, and posterity was the last thing on their minds.

The NAS archives, which were of importance to the prologue; chapters two, three, and eleven; and to parts of the other sections have been well mined, and others have written extensively of the IGY and its relationship to the subsequent development of space science in the U.S. My "angle" was to explore the same material for the seeds of space technology and of application satellites.

Both oral and written primary sources are of equal importance to the navigation section. The pre-Transit chapters were possible only because of long and repeated interviews, while the chapters on Transit were possible only because of the material in APL's archives.

The meteorology section is based on interviews, a few primary sources, and secondary sources. It provides the clearest example of the emergence of application satellites from the IGY. But access to declassified primary sources will eventually make the history of meteorology satellites much more complete.

The communication section is the most heavily based on primary source written records, supplemented with a few interviews.

Prologue

The primary source of material for the prologue (also for chapters two, three, and eleven) is the archival material about the International Geophysical Year stored at the National Academy of Sciences in Washington, DC.

Of particular importance were minutes of the USNC Committee of the IGY; minutes of the Executive Committee of the IGY, and the minutes of the Technical Panel on Rocketry. The account of James Van Allen's dinner party (page 1) comes from an oral history given by Dr. Van Allen to David DeVorkin in February, June, July, and August of 1981 for the National Air and Space Museum.

Observations about Lloyd Berkner's character were pieced together from impressions gained by reading minutes of IGY committee meetings (page 1). His name crops up in records for communications and meteorology satellites and in the development of early U.S. space policy. The biographical files at the NASA History Office describe a naval officer who, when he died, was buried with full military honors and someone who opposed scientific secrecy. Besides being the originator of the idea for the IGY, Berkner was president of the International Council of Scientific Unions.

Drawer 1 of the archives of the National Academy of Sciences contains program proposals for the IGY, including one from Paul Siple highlighting concerns then felt about global warming (page 3).

Drawer 2 of the NAS archives contains the minutes of the first meeting of the U.S. National Committee for the IGY held 26–27 March 1953. Also in drawer 2 are to be found tentative proposals for the IGY 1957–1958 prepared by the USNC for the IGY, 13 May 1953 (page 3).

The anecdote that administration officials said, "Joe, go home," was related by Kaplan himself in a speech to mark the tenth anniversary of Explorer (page 4). A copy of the speech was among Verner Suomi's papers.

The account of what happened in Rome in 1954 and the budget figures for the IGY are found in *Vanguard—A History,* by Constance Green and

Milton Lomask (page 4). The book is part of the NASA History Series, SP4202.

Information about President Eisenhower's intelligence needs (p. 4–5) and his national security policy comes mainly from ... *the Heavens and the Earth: A Political History of the Space Age,* by Walter McDougall (Basic Books, 1985). This book addresses what has been the central mystery of the U.S. space program—why the Eisenhower administration chose the Vanguard rather than the Explorer program for the development of the first U.S. satellite.

The existence of the Killian panel is well known, and its existence is written about in numerous accounts of the time, but McDougall's discussion is the most exhaustive I encountered (page 4).

The most detailed and up-to-date information about the Killian panel (page 4) and its influence on the Eisenhower administration's policy and of the way that national security considerations impacted the development of the IGY are to be found in R. Cargill Hall's article "The Eisenhower Administration and the Cold War, Framing American Astronautics to Serve National Security," in *Prologue, Quarterly of the National Archives.*

The exact sequence of events in which Donald Quarles, assistant secretary of defense for research and development, approached senior scientists of the IGY is not clear when one looks at Cargill Hall's article and the sequence of events that surrounded planning of the IGY (page 5). However, minutes of the IGY suggest that in the light of Cargill Hall's article, some senior scientists other than Joseph Kaplan knew or guessed the national security agenda that necessitated developing a satellite with a largely civilian flavor.

The first meeting of the USNC of the IGY was chaired by Joseph Kaplan (page 3).

During the third meeting on November 5–6, 1954, James Van Allen commented on the usefulness of rocketry studies. At this time, though various international bodies had endorsed the idea of a satellite program forming part of the IGY, Van Allen's presentation referred to sounding rockets, i.e., those that carry instruments aloft but fall back to Earth without entering orbit.

During the fourth meeting, on January 14 and 15, 1955, Harold Wexler spoke of gaps in the meteorological data, and Homer Newell, in the absence of James Van Allen, told the committee that the sounding rocketry work would be undertaken entirely by the agencies of the Department of Defense, provided that the National Science Foundation secured the necessary funding from Congress.

By this time, much of the debate concerning the importance to the United States of adopting a satellite program as part of the IGY had moved to the USNC's executive committee and to a working group of the technical panel on rocketry (see notes and sources for chapter three) as well as to the National Security Council.

Reports on Leonid Sedov's announcement at the sixth meeting of the International Astronautical Federation in Copenhagen appeared in the *Baltimore Sun* as well as in other newspapers, dateline August 2, 1955 (page 6).

Chapter one: New Moon

Chapter one, inasmuch as it refers to Sergei Korolev, is based on secondary sources. Anyone interested in an account based on primary sources should look for a biography by Jim Harford that was published this fall (1997) by John Wiley and Sons.

Even the secondary sources about Sergei Korolev are sparse, and in each case it was essential to consider carefully who wrote it, where the author was at the time, and when and where the account of Korolev was published. I also attempted to establish whether any given anecdote had similar sources or whether it came from genuinely independent accounts. I have allowed my imagination to have more play in this chapter than in the rest of the book.

Despite being the chief designer of cosmic rocket systems, Korolev was unknown in the West at the time of the launch of *Sputnik* (page 8). In a 1959 bibliography on Soviet missiles and state personnel (Library of Congress reference TL 789.8.R9H21) intended to give the U.S. technical world information about Soviet activities, Korolev is listed in a publication from HRB-Singer only as someone interested in liquid-fueled rocket engines.

Grigory Tokady, a defector, first disclosed that Korolev was the chief designer for the Soviet space program during a meeting of the British Interplanetary Society in 1961. His revelation was not widely reported.

One of the best accounts of Sergei Korolev's early life is by Yaroslav Golovanov, *Sergei Korolev: The Apprenticeship of a Space Pioneer* (Novosti, 1976). The book gives a brief account of the launch of *Sputnik,* but otherwise is devoted entirely to Korolev's youth; his early poetic efforts; and his relationships with his mother, grandmother, and stepfather (pages 15 and 16). It is the only book I found that explores the events and relationships that shaped the man. Clearly, the author has interviewed many people who knew Korolev and has tried to evaluate some of the folklore that has grown up around him, in particular, Korolev's purported meeting with Tsiolkovsky. Irritatingly, Golovanov's account stops before Korolev was arrested by Stalin's secret police. The author is reported to have completed a full biography, written in Russian, and to be in search of a publisher.

Details of Korolev's state of mind in prison in Moscow, his activities there, and the impact that his incarceration in Kolyma had on him appear in Georgii Oserov's book, published in Paris, *En Prison avec Tupolev* (A. Michel, 1973). Oserov was in prison with Korolev and the elite of national aeronautics.

Walter McDougall's book . . . *the Heavens and the Earth: A Political History of the Space Age* chronicles the USSR's fascination from Lenin's time with technology and the country's national goal of achieving technical supremacy. This goal led to internal tensions and confrontations between the government and the intelligentsia.

McDougall describes how Marshal Tukhachevsky became a victim of Stalin's purges and how in 1938 the rocketeers, including Korolev, joined Stalin's earlier victims, the aircraft designers, in the Gulag's prison camps.

McDougall reports that Korolev's failures in early 1957 encouraged his rival Chalomei to attempt to have him dismissed.

Other less detailed accounts of Korolev's early years exist. *The Kremlin and the Cosmos,* by Nicholas Daniloff (Knopf, 1972), for example, provides a good summary of Korolev's schooling without the attempts that Golovanov makes to explore his psyche. The account is hopelessly inadequate once one enters the difficult years of arrest, concentration camps, and divorce. Daniloff does, however, mention briefly that there were "trying and despairing situations" in Korolev's life.

Daniloff also gives an account of the launch of *Sputnik* and of the engineers retiring to an observation bunker a kilometer from the launch pad (pages 10, 18, and 19).

Aleksei Ivanov, an engineer who worked on *Sputnik,* also recounts the launch (pages 10, 18, and 19) in an article in *Isvestia* marking the tenth anniversary of *Sputnik.* He wrote, "I watch not moving my eyes away, fearing to blink so as not to miss the moment of liftoff."

Another book, more a hagiography than a biography, about Korolev is *Spacecraft Designer: The Story of Sergei Korolev* (Novosti, 1976). The author, Alexander Romanov, says that he first met Korolev in 1961. If one is careful, some details seem worth extracting from this book. The author describes Korolev as a heavyset man, a description that photographs support. His account of Korolev's small, wood-paneled office with blackboard, chalk, lunar globe, bronze bust of Lenin, and model of *Sputnik* seems plausible, as does his account of a formidable intellect and an energetic man with willpower, energy, and vision.

Romanov repeats uncritically the story that Korolev met Konstantin Tsiolkovsky in Kaluga in 1929. Romanov reports that after that meeting, Korolev said, "The meaning of my life came down to one thing—to reach the stars."

Romanov demonstrates Korolev's dedication to rocketry with an extract from a letter that Korolev wrote to his second wife in which he wrote, "The boundless book of knowledge and life . . . is being leafed through for the first time by us here."

Romanov also reports that it was Korolev who wanted *Sputnik* to be spherical, which, given Korolev's authority in the program, seems likely. Romanov says that Korolev said, "It seemed to me that the first *Sputnik* must have a simple and expressive form close to the shape of celestial bodies."

More details of Korolev's character—his strictness, compassion, and demanding nature—appear in a collection of essays entitled *Pioneers of Space,* which were compiled by Victor Mitroshenkov (Progress Publishers, 1989). Korolev's engineering intuitiveness apparently amazed his colleagues.

In one of the essays, Nikolai Kuznetsov, who headed the cosmonaut training center from 1963, wrote that Korolev liked the cosmonauts to meet the ground staff so that "cosmodrome specialist and cosmonaut could look one another in the eye." It was Korolev's way of ensuring that work on Earth was carried out conscientiously. This, together with his recorded friendship with cosmonauts Yuri Gagarin and Alexei Leonov, is the basis for my saying on page 8 that Korolev cared deeply about the fate of his cosmonauts.

Another essay by Pavel Popovich and Alexander Nemov says that people found Korolev either sincere, unpretentious, and accessible, or mercilessly strict and demanding with slackers. He was, they say, intolerant of vanity.

The essays include brief accounts of the months before the launch and contain nice details, such as Korolev's habit of lifting his little finger to his eyebrow when vexed.

Other books in which snippets of information about *Sputnik,* Korolev, and the space race appear that back up information from the main sources include *Soviet Rocketry, Past, Present and Future,* by Michael Stoiko (Holt, Rinehart and Winston, 1970); *Russians in Space,* by Evgeny Riabchikov (prepared by Novosti Press Agency, published New York, Doubleday, 1971); *Soviet Writings on Earth Satellites and Space Travel,* editor Ari Sternfield (Freeport, NY, Books for Libraries Press, 1970); *Red Star in Orbit,* by James Oberg (Random House, 1981). Oberg quotes Solzhenitsyn as saying that Korolev worked on his rocket at night; *The Sputnik Crisis and Early United States Space Policy,* by Rip Bulkeley (Indiana University Press, 1991), and *Race into Space: The Soviet Space Program,* by Brian Harvey (Ellis Horwood, a division of John Wiley, 1988).

A description of the location of the Baikonur cosmodrome and the relative position of Korolev's cottage can be found in this 1986 edition of *Jane's Spaceflight Directory.*

Information about the events of the IGY meeting on rockets and satellites in Washington, DC, appears in the archives of the National Academy of Sciences.

The IGY meeting in Washington was reported in the *New York Times,* October 4, 1957.

The anecdote about Korolev's conversation with Alexei Leonov a few nights before he died comes from Jim Harford. Harford also talked to me about Korolev's visit to Peenemünde after World War II.

Khrushchev's views on the significance to the Soviet Union of ICBMs are to be found in his autobiography, *Khrushchev Remembers: The Last Testament,* translated by Strobe Talbot (Little Brown).

Khrushchev describes his casual attitude toward Korolev's news of the launch of *Sputnik* to James Reston in an interview published in the *New York Times* on October 8, 1957. Khrushchev says he congratulated Korolev, then went to bed.

An understanding of what life in prison was like for Korolev can be found in *The First Circle,* by Aleksandor Solzhenitsyn.

Though chapter one is about Korolev because it was his satellite that opened the space age, Robert Goddard was the man who built and launched the first liquid-fueled rocket—a fact of which Korolev was well aware. A companion book for anyone interested in the pioneering days of rocketry therefore is *Robert H. Goddard: Pioneer of Space Research,* by Milton Lehman (Da Capo Press, 1988). The footnote about the launch of the world's first liquid-fueled rocket comes from this book. Lehman's book was published first as *This High Man* (Da Capo Press, 1963).

I found information about general historical events, such as the coup that Khrushchev faced down in June 1957 (page 11), in *A History of the Soviet Union,* by Geoffrey Hosking (Fontana, 1985).

Chapter two: Cocktails and the Blues

A flavor of the times described in this chapter comes from my interviews with William Pickering, Milton Rosen (technical director of Project Vanguard), John Townsend, and Herbert Friedman (of the Naval Research Laboratory and a member of the USNC).

When the Soviet embassy's party began (page 21) on the evening of October 4, 1957, did Anatoli Blagonravov know that *Sputnik* had been launched? William Pickering thinks not. John Townsend doesn't know, but he says that Homer Newell (scientific program coordinator for Project Vanguard), who was with them, was convinced that Blagonravov did know.

My physical description of Blagonravov comes from reports in the *New York Times* during the week of the conference (page 22). I'm guessing that he drank vodka.

My description of William Pickering is from my own observations (page 22).

Information about Vanguard came mainly from my interview with Milton Rosen and was bolstered by articles and papers he gave me, including one by John Pierce describing how satellites might be used for telecommunications, one that Rosen sent on 3 March 1955 describing the utility of the Viking rocket for a satellite launch vehicle, and a memo on the same date from John Mengel and Roger Eastman outlining Minitrack.

A considerable amount of useful information about Vanguard can also be found in Green's and Lomask's book and in the NASA History Office (see notes for the prologue). Green and Lomask give details of the costs and of the rival merits of the Projects Orbiter and Vanguard proposals, including details of the miniaturization to be found in Vanguard (page 26).

Milton Rosen told me that von Braun asked for a chance to make a second presentation of Project Orbiter to the Stewart committee, and of the anxiety with which he (Rosen) watched the presentation and the disbelief and elation felt by the NRL team when they learned that the committee had backed Project Vanguard (page 26).

My interview with William Pickering and his oral history, given to Mary Terral for the archives of the California Institute of Technology, provide details of Project Orbiter and how it evolved from von Braun's original proposal.

Information about the long playing rocket (the euphemism by which a rocket capable of reaching orbit was known), its costs, and the costs of the satellite program, as well as their acceptance by the USNC–IGY is found in the following minutes, located at the National Academy of Sciences: third meeting of the USNC executive committee (January 7, 1955), during which the technical panel on rocketry was asked to report on the technical feasibility of satellites; first meeting of the technical panel on rocketry (January 22, 1955) during which a subcommittee comprising William Pickering, Milton Rosen, and John Townsend was formed; first meeting of the subcommittee evaluating the feasibility of a satellite launch (February 3 and 4, 1955, in Pasadena). No minutes, though William Pickering remembers the first meeting. Fourth meeting of the executive committee of the USNC, during which members were told that the technical evaluation of a satellite was ongoing; On March 5, 1955, there was a meeting between Joseph Kaplan and Hugh Odishaw, the administrative secretary of the USNC, to discuss security procedures surrounding the work of the subcommittee of the technical panel on rocketry. They concluded that the report would be classified but that the committee would prepare an unclassified report for the executive committee; On March 9, 1955, the technical panel on rocketry accepted a classified report on the feasibility of launching a satellite and prepared an unclassified version for the USNC's executive committee; On March 8–10, the fifth meeting of the USNC executive committee discussed the rocketry panel's report and debated whether to back the inclusion of satellites in the IGY. Unusually, the notes from this meeting are handwritten and hard to decipher. The most vocal discussants were Merle Tuve and Athelstan Spilhaus. Tuve expressed doubt about the inclusion of a satellite in the IGY; Spilhaus was strongly in favor. What caused Tuve concern was the classified nature of the project. In the end, the meeting agreed that if the long playing rocket were still classified by January 1956, then it should be dropped from the program. The seventh meeting of the USNC took place on May 5, 1955. An agenda item on the LPR refers to an attachment that is not included in the archives. This meeting was in the period following Quarles' approach to the IGY and prior to Eisenhower's public announcement that there would be a satellite program. Hugh Odishaw was writing and phoning the NSF at this stage, pushing for a decision on the satellite budget and seemingly unaware of the higher policy decisions in which the satellite program was caught.

Odishaw's letters and Joseph Kaplan's correspondence with Alan Waterman, the director of the National Science Foundation, and Detlev Bronk, the president of the National Academy of Sciences, are in the correspondence files of the IGY archives at the NAS.

Details of the curtailment of the satellite program (page 30) in the year following President Eisenhower's announcement that it would go ahead are also to be found in the archives of the NAS. The USNC discussed the Earth satellite program at its tenth meeting on 13 July 1956. The minutes, classified as administratively confidential, say, "For reasons of economy, the Earth satellite program has been curtailed from 12 attempted launching to six." Hugh Odishaw drew the committee's attention to the need for confidentiality, "lest knowledge of the curtailment of the program should lead to an international loss of prestige by the U.S."

The interaction between the IGY and national security policy comes primarily from R. Cargill Hall, "The Eisenhower Administration and the Cold War, Framing American Astronautics to Serve National Security," in *Prologue, Quarterly of the National Archives,* spring 1996. It seems likely that the criteria that the Stewart Committee were given in order to decide between Projects Orbiter and Vanguard were chosen to ensure that the IGY could indeed be a "stalking horse" for the launch of a reconnaissance satellite.

Information about the Stewart Committee (page 25) can be found in the oral history by Homer Joe Stewart in the archives of the California Institute of Technology and in *Vanguard—A History* (NASA History Series SP4202), by Constance Green and Milton Lomask. Milton Rosen also gave me information about the way the Stewart Committee voted and the reasoning behind their decision.

Homer Joe Stewart was interviewed by John L. Greenberg on October 13 and 19 and November 2 and 9, 1982, for an oral history, which is in the archives of the California Institute of Technology.

Details of how the scientists learned of the launch of *Sputnik* come from conversations with Walter Sullivan, William Pickering, and John Townsend (page 30).

Books consulted for chapter 2, as well as for the prologue and chapter 11 are *Beyond the Atmosphere, Early Years of Space Science,* by Homer E. Newell (SP4211—NASA History Series); *Science with a Vengeance, How the Military Created the US Space Sciences after World War II,* by David DeVorkin (Springer-Verlag, 1992); *The Viking Rocket Story,* by Milton W. Rosen (Faber, 1955).

Chapter three: Follow That Moon

William Pickering's state of mind and actions following Lloyd Berkner's toast to the Soviets come from my interview with him. He described also the error in calculation they had made and the phone calls that poured into the headquarters of the IGY (pages 30–34).

Information about Project Moonwatch comes from my interviews with Roger Harvey, Henry Fliegel, and Florence Hazeltine.

Information on the radio tracking program comes from interviews by Green and Lomask with Daniel Mazur and Joseph Siry in the NASA History Office, as well as from the following papers: John T. Mengel, "Tracking the Earth Satellite, and Data Transmission by Radio," *Proceedings of the IRE* (44), 6, June 1956; John T. Mengel and Paul Hergert, "Tracking Satellites by Radio," *Scientific American* (198), 1, January 1958.

Information about the goals of the IGY satellite program and details of the optical and radio tracking systems and the technical and budgetary difficulties faced comes from minutes of the IGY committees, subcommittees, panels, and working groups:

Minutes of the first meeting of the Technical Panel on the Earth Satellite program (TPESP), October 20, 1955. At this meeting the panel defined the program's goals (page 32).

10 November 1955: An ad hoc meeting of the technical panel on Earth satellites (TPESP) convened to discuss the budget for the program, which had to be ready for a presentation to Congress and the Bureau of the Budget (predecessor to the current Office of Management and Budget) by March 1956. Homer Newell said that important things to be budgeted for were radio and optical tracking and scientific instrumentation. The NRL, who were the experts at radio tracking, wanted stations distributed between latitudes of 35 degrees north and south of the equator. The TPESP wanted to add two more tracking stations to extend coverage to 45 degrees. These tracking stations eventually became known as minitrack.

The optical tracking program was discussed in greater detail at the second meeting of the TPESP, on November 21, 1955. Fred Whipple, director of the Smithsonian Astrophysical Observatory, presented a report prepared by himself and Layman Spitzer. The TPESP recommended that up to $50,000 be awarded to the SAO immediately to set up a series of observing stations. At the time, Whipple's proposal was for twelve observing stations and an administrative and computer analysis center. He also called for collaboration with amateur observers.

During the third meeting of the TPESP, on January 28, 1956, the difficulties of tracking began to emerge. A letter from Homer Newell on the problems of visual and photographic tracking of Earth satellites was read. It was not known whether radio tracking would work (see page 36). The expectation at the time was that there was only a fifty percent likelihood of minitrack succeeding; hence the need for optical tracking.

27 June 1957: The twelfth meeting of the USNC pointed out that there were still problems with the tracking system.

At the seventh meeting of the TPESP on September 5, 1956, John Hagan and Fred Whipple respectively updated the panel on radio and optical tracking. By now, Whipple had made contact with amateurs in an attempt to improve the chances of acquiring the satellite optically. The army, for example, had four hundred binocular elbow telescopes that volunteers, like Florence Hazeltine, could use at military bases.

The twelfth meeting of the TPESP, on October 3, 1957, the eve of the launch of *Sputnik,* opened with a discussion about how to track a Russian satellite. Fred Whipple explained delays in development of the cameras for optical tracking. It was during this meeting that the delays in delivery of the cameras prompted Richard Porter to say, "I have a number of times threatened to go up to Stanford and beat on tables.... Fred [Whipple] has so far frankly discouraged my doing so."

At the thirteenth meeting of the TPESP, on October 22, 1957, it was reported that delivery of optics from Perkin Elmer had been increased and brought forward.

The fifteenth meeting of the TPESP, on January 7, 1958, demonstrates the poverty of information about the *Sputniks'* orbits. Whipple said, "We've not had a scrap of radio information." Richard Porter, who headed the panel, said, "We may have underestimated again the difficulty of tracking and photography." Pickering said, "The Soviet thing caught everyone off base" (page 34).

That the Soviets were also conducting the same basic science experiments and were interested in ionospheric refraction, tracking, and propagation effects comes from *Selected Translations from Soviet-Bloc International Geophysical Year Literature. Artificial Earth Satellite Observations* (New York, U.S. Joint Publication Research Services, 1959) and *Selected Reports Presented by the USSR at the Fifth Meeting of the Special Committee for the International Geophysical Year* (New York, U.S. Joint Publication Research Services, 1958).

Details of the optical tracking program can be found in the annual reports of the SAO for 1961 and 1963.

Green and Lomask (*Vanguard—A History,* NASA History series SP4202) describe John Mengel's actions when *Sputnik* was launched (page 35).

Navigation section

Individuals interviewed for the navigation section are as follows:

Bob Danchik★ (Transit's penultimate project manager), Bill Guier★ (physics), George Weiffenbach★ (physics), Lee Pryor★ (software development and Transit's last project manager), Carl Bostrom★ (physics and later the director of APL), Henry Elliott★ (antennas), Lee Dubois★ (command, control, and tracking), Charles Pollow★ (assistant program manager), Laurence Rueger★ (time and frequency systems), Tom Stansill★ (receivers), Russ Bauer★ (software), Charles Bitterli★ (software), Harold Black★ (physics/orbital mechanics), Ben Elder★ (memory designer), Eugene Kylie★ (receivers), Barry Oakes★ (rf systems), Charles Owen★ (mechanical design), Henry Riblet★ (antenna design on Transit), Ed Westerfield★ (receiver design), John O'Keefe (satellite geodesist), Gary Weir (naval historian), Commander William Craft (commander and director of seamanship at the U.S. Naval Academy, in Annapolis), Brad Parkinson (GPS project manager), Group Captain David Broughton (director of the Royal Institute of Navigation), and Dave Smith (satellite geodesist, currently at the Goddard Space Flight Center, Greenbelt, Maryland).

An asterisk denotes that an individual was a member of the Transit team.

Some of the above were interviewed in great depth and over many hours, weeks, and in the case of Guier and Weiffenbach, months; a very few spoke to me for as little as half an hour.

Chapter five: Polaris and Transit

Information about Polaris (pp. 49 and 51 – 53), its purpose and development, emerged during many hours of interviews with members of the Transit team.

Books that provided background for chapter five include *The Polaris System Development: Bureaucratic and Programmatic Success in Government,* by Harvey M. Sapolsky (Harvard University Press, 1972); *Forged in War: The Naval–Industrial Complex and American Submarine Construction, 1940–1961,* by Gary Weir (The Naval Historical Center, 1993).

The potted history of navigation in this chapter (pages 50 and 51) draws on interviews with Commander William Craft and Group Captain David

Broughton, and on *From Sails to Satellites: The Origin and Development of Navigational Science,* by J.E.D. Williams (Oxford University Press, 1992).

Transit's status as brickbat-01 and the meaning of this terminology (page 53) emerged during interviews with Transit team members.

The fact that radars capable of detecting a periscope's wake were being developed at the end of the 1950s and in the early 1960s (page 54) comes from George Weiffenbach.

How the technology of the Transit receivers evolved (page 55) comes from Tom Stansill and from papers and old sale brochures that he sent to me.

The history of APL (page 56) comes from *The First 40 Years, JHU APL* (Johns Hopkins University Press, 1983).

Notes about Frank McClure's and Richard Kershner's previous careers (page 56) come from *The First 40 Years, JHU APL* and from press releases and briefing papers sent to me by Helen Worth, the APL's press officer.

The fact that Ralph Gibson was sounded out as a potential first director of the Defense Research Establishment is mentioned by Admiral William Raborn, head of the Fleet Ballistic Missile program until 1962, in an oral history at the Naval Historical Center, in Washington, DC (page 57).

Chapter six: Heady Days

The contents of the National Security Council's agenda and mention of Lay's phone calls to Alan Waterman (page 58) are among Waterman's papers in the Library of Congress.

Information about Tycho Brahe and Johannes Kepler (page 59) comes principally from an essay on Kepler by Sir Oliver Lodge in *The World of Mathematics,* edited by James R. Newton (Tempus, 1956). Textbooks consulted for pages 60 and 61 are *Introduction to Space: The Science of Spaceflight,* by Thomas D. Damon. A foundation series book, chapter three deals with orbits (Orbit Book Company, 1989); *The Feynman Lectures on Physics,* volume one, chapter 7 (Addison Wesley, 1963).

Bill Guier and George Weiffenbach supplied the information for pages 61 to 65.

The textbook consulted for pages 65 to 69 is *The Feynman Lectures on Physics,* volume one, chapter 34 (Addison Wesley, 1963).

The material on pages 70 to 72 is based on the memories of George Weiffenbach, Bill Guier, and Henry Elliott. They all spoke to me independently. Guier and Weiffenbach spoke to me many times. Each remembered things a little differently, but their memories differed little in substance. These memories are, to my knowledge, the only sources of information for Guier's and Weiffenbach's work in October at APL. As far as I know, there are no written records, not even laboratory notes, that support the assertion that Guier's and Weiffenbach's work was unofficial and an indulgence of curiosity.

Chapter seven: Pursuit of Orbit

The main sources of information for this chapter are Bill Guier and George Weiffenbach.

The problems in putting this chapter together were that there are no primary written sources that directly confirm what Guier and Weiffenbach did and when, and they have both told the story several times, including on tape in a 1992 APL video. Thus it took some time to recall memories.

The only way around this seemed to me to go over and over the same ground from as many different angles as possible. And both Guier and Weiffenbach seemed to take to this approach as the proverbial duck takes to water. Each time I gleaned another fact, no matter how small, from one of their colleagues or from a published paper, I went back to one or another of them to ask more detailed questions or the same questions in a different guise. As I learned a little more about the physics for myself, I also went back to them.

The result is chapter seven, which is corroborated as much as possible by memories from other people.

Certain bounds to the time when their work was done are set by undisputed dates, such as the launch of *Sputnik II* and the day that their work became an official project.

Lee Pryor, who was at that time studying computing at Pennsylvania State University, confirms much of what Guier says about coding for the Univac.

The richness of information available on the Doppler curve (page 75) is apparent in a highly mathematical in-house paper (Part of APL's Bumble-bee series) "Theoretical Analysis of the Doppler Radio Signals from Earth Satellites" published in April 1958.

Charles Bitterli remembers working on an algorithm for least squares (page 78).

Henry Elliott's memories corroborate Weiffenbach's view of himself as a painstaking researcher who would check the quality of data in detail (page 79).

Chapter eight: From *Sputnik II* to Transit

Project D-54, to determine a satellite orbit from Doppler data, APL archives (page 82).

Guier and Weiffenbach's briefing about their work (page 82) is from my interview with Harold Black.

Information about Guier's and Weiffenbach's early work on the ionosphere (page 82) is from interviews with Weiffenbach and Guier.

A textbook consulted on ionospheric refraction is *The Feynman Lectures on Physics,* volume one, chapter 28 (Addison Wesley, 1963).

Weiffenbach's memo to Richard Kershner and the first Transit proposal (pages 84 and 85) are in the archives of the Applied Physics Laboratory.

Henry Riblet told me of the need to modify the design of circularly polarized transmitters for Transit's spherical surface (page 85).

Information about O'Keefe and his views (pages 85 and 86) came from my interview with John O'Keefe.

Views about Frank McClure's character (page 87) came from nearly every member of the Transit team that I interviewed.

Frank McClure's ideas for a navigation satellite are in a memo dated March 18, 1958, reproduced in *The First 40 Years, JHU APL* (Johns Hopkins University Press).

What McClure said to Guier and Weiffenbach about his satellite navigation idea is a story that both told me separately (page 88).

Chapter nine: Kersher's Roulette

Comments about Richard Kershner (page 91), his approach to the job of team leader and to engineering, are based on the views of different Transit team members.

The First Transit Proposal, 4 April 1958, (APL Archives) gives details of the satellite and incorrectly suggests that the ionosphere might be the biggest problem facing satellite navigation (pages 92–94).

Limits on orbital configuration and its relationship to ground stations (page 94) are from interviews with Guier and Weiffenbach.

The section in this chapter on the search for longitude had a number of secondary sources:
　　John Harrison: The Man who Found Longitude, by Humphry Quill (Baker, 1966).
　　History of the Invention by John Harrison of the Marine Chronometer, by Samuel Smiles (Press Print).
　　Memoirs of a trait in the character of George III of these United Kingdoms, by John Harrison (W Edwards, 1835).
　　John Harrison and The problem of Longitude, by Heather and Mervyn Hobden (Cosmic Elk, 1989).
　　"The Longitude," an essay by Lloyd A. Brown in volume two of *The World of Mathematics,* edited by James R. Newman (Tempus, 1956).

Kershner's trips to the Pentagon (page 97) were remembered by both Guier and Weiffenbach. Though there is no written record of these trips at APL, he presumably had to go back and forth several times.

Transit on Discovery is mentioned several times in memos, letters, and progress reports of Transit in the APL archives (page 98), and various members of the team explained that it was part of DoD efforts to determine Earth's gravitational field and thus, of course, the forces that would act on a ballistic missile in flight.

The details in pages 98 to 104 were extracted from numerous reports and memos in the APL archives, from interviews with the Transit team members, and from memos and papers that Henry Elliott had kept.

Chapter ten: The Realities of Space Exploration

The account of the launch of Transit 1A is pieced together from comments from different team members. Lee DuBois, for example, remembered the tears in his eyes when the satellite failed (page 105).

Details of the twenty-five-minute flight and the results gleaned come from the records that Henry Elliott had kept and from reports in APL's archives (page 106).

Numerous memos and reports in the APL archives testify to Kershner's industry in preparing for Transit 1B and Transit 2A (pages 106 and 107).

John Hamblen's undated, typed note requesting the team members to document component testing (page 107) is among the papers in the APL archives.

My description of how the launch might have been is pieced together from people's memories and photographs of later launches found at APL.

A progress report details what happened scientifically following the successful launch of Transit 1B (page 108–109). Bill Guier explained what the report meant and supplemented it with his own memories.

A textbook consulted on the geoid (page 110) is *Theory of the Earth,* by Don L. Anderson (Blackwell Scientific Publications, 1989).

Information about how APL's understanding of the geoid developed comes from papers and from interviews with Bill Guier and Harold Black (page 112).

Dave Smith, director of the division of terrestrial physics at the Goddard Space Flight Center gave me some very basic understanding of satellite geodesy.

Information about developing subsequent orbital determination programs and about the computers comes from reports in APL's archives and from interviews with Harold Black and Lee Pryor (pages 113 and 114).

Information about the problems with the solar cells (page 114) came from Bob Danchik.

General comments on satellite navigation

1. At different places in this section, I have alluded to alternative ideas for navigation satellites. One, explained in "Navigation by Satellite" in *Missiles and Rockets* in October 1956, even talks of utilizing the Doppler shift for a navigation satellite. But this paper envisages almanacs and tables of position and calculations of the distance at closest approach. It implicitly assumes that the orbit would be known and, inevitably because it was written in 1956, does not account for the impact on orbits of Earth's complex gravitational field, nor for the impact of the ionosphere on the received signal. The paper does envisage the use of computers, but not the sophisticated curve-fitting techniques of Guier and Weiffenbach.

2. "Possible Use of Syncom as a Navigation System—Microwave Loran." Memo for files, From L.M. Field cc L.A. Hyland (HAC archives 1990-09 box 6 folder 22).

This memo argues that Syncom would make a better navigation satellite than Transit if the station keeping were adequate. It expands on a memo by Donald Williams (see communications section) written on September 1, 1959.

Chapter 11: Move Over, Sputnik

Khrushchev's belligerent attitude toward the United States, which presumably set the tone for the excerpt from *Soviet Fleet* (page 119), is clear in an interview with James Reston for the *New York Times* that appeared on October 9, 1957. In the interview, Khrushchev said that the U.S. was causing all the trouble [between the two countries] because it negotiated with the USSR as if it were weak. Khrushchev told Reston that the USSR had all kinds of rockets for modern war and spelled out the fact that since the USSR could launch a satellite, it had the technology for intercontinental ballistic missiles. In his interview, Khrushchev uses the terms "imperialist warmongers" and "reactionary bourgeoisie" when speaking of the U.S.

Eisenhower's attitude to this rhetoric is apparent in his State of the Union message on January 9, 1958. He said, "The threat to our safety, and to the hope of a peaceful world can be simply stated. It is communist imperialism. This threat is not something imagined by critics of the Soviets. Soviet spokesmen, from the beginning, have publicly and frequently declared their aim to expand their power, one way or another, throughout the world."

The failure of the launch of *Vanguard* on December 6, 1957, (page 120) is retold in Green's and Lomask's book in the NASA History Series, *Vanguard, A History.*

General Medaris's call to JPL et al. (page 121) is discussed by William Pickering in his oral history in the archives of the California Institute of Technology.

Pickering's attempts to contact James Van Allen are discussed both in Pickering's oral history at Caltech and in James Van Allen's oral history in the National Air and Space Museum (page 121–122).

In the oral history and in his interview with me, Pickering described the journalist who tracked him down in New York immediately prior to the launch.

The account of the launch of *Explorer 1* (pages 122–123) comes from the oral histories of Pickering and Van Allen, from Pickering's interviews with me, from the Green and Lomask book (*Vanguard—A History*) and from accounts in the *New York Times.*

James Van Allen describes his isolation and concerns on the launch of *Sputnik* in his oral history at the National Air and Space Museum.

In his oral history and when talking to me, Pickering describes the wait for the acquisition of *Explorer 1* and his trip to the NAS to face the world's press once the satellite was in orbit.

Details on pages 123–125 are a synthesis of extracts from minutes of the USNC for the IGY, the executive committee of the USNC, the TPESP, the Technical Panel on Rocketry and the Working Group on Internal Instrumentation.

An ad hoc meeting of the TPESP, November 10, 1955, discussed the budget for the whole satellite program. Homer Newell emphasized the importance of allocating money quickly for optical and radio tracking and for scientific instrumentation.

The need to understand the organizational relationship between the different official bodies was mentioned. This was an early warning of a struggle for control of the program. The NRL was emphatic that it retain control of the launch vehicle and to a large extent over the scientific instrumentation. A formal statement said, "The NRL desires the advice and guidance of the USNC with respect to scientific instruments. . . . advice will be followed in so far as is possible without compromising the achievement of a first successful launch."

The question of organizational relationships and responsibility resurfaced nearly a year later. In an attachment to the minutes of the seventh meeting of the TPESP on September 5, 1956, a note records an informal meeting at the Cosmos Club between Richard Porter and Admiral Bennet during which there was discussion about whether the satellite program was a DOD program with IGY participation or vice versa.

The third meeting of the TPESP took place in the Founders' Room at the University of Michigan. The panel discussed the possibility that the

Soviets might launch a satellite in 1956 and agreed to make formal enquiries through the CSAGI, the international organizing body for the IGY.

During the third meeting of the TPESP, the chair, Richard Porter, appointed James Van Allen to head a working group on internal instrumentation. The panel's job was to review the proposals for experiments submitted to the panel.

At the fifth meeting of the TPESP on April 20, 1956, Homer Newell said that a master schedule for the development of Vanguard was being drawn up and would be circulated to the TPESP as soon as it was ready. He described several satellite configurations and reported on development of the rocket's first stage. During the same meeting there was discussion of budget overruns.

During the meeting, Richard Porter also discussed the concerns that there might be only one satellite launched. He said, "... if the plan really is to stop after you get one good one, then we had best discontinue most of the work of this panel. In fact, my own feeling is that the program would not be worth doing if this were the intent." The DOD observer replied, "It is the feeling of the executive branch of government that our present job is to get one up there, and it is most unlikely ... that we will answer in any other way than to say it is a future not present decision [how many satellites are launched] (page 124).

The meeting went on to question whether they had a right to spend taxpayers' money on a tracking system when there might only be one satellite.

In an attempt to justify six launch attempts, Dr. Porter said, "Actually, the six rides was determined by another group headed by Dr. Stewart as being the minimum that ought to be fired to get one good one. This is the best guess of some guided missile experts."

Porter argued that there should be more ideas put forward for satellite experiments because he believed there would be an extended program of satellite launches and that once in place, the tracking system would be cheaper to operate than it was to set up.

Meteorology section

During two trips to Wisconsin in the summer of 1992, I spent many hours interviewing Verner Suomi. He provided a lot of background and color to the early story of meteorology satellites. True to his experimenter's approach to life, he was at the time trying a novel therapy following three heart operations. He had the same degree of curiosity about the experiment he was participating in as he had in his meteorological work.

Others interviewed for this section include
Dave Johnson, Robert White, Joseph Smagorinsky, Pierre Morel, P. Krishna Rao, Bob Sheets, Leo Skille, Bob Sutton, and Bob Ohckers.

When I interviewed P. Krishna Rao and Bob Sheets, I was considering writing a book that brought the story of meteorology satellites right up to date. In the end, that wasn't possible but these interviews helped give me a sense of the evolution of the technology, and what I learned from them is, I hope, implicitly present in this section.

Chapter twelve: A Time of Turbulence

The promise of satellites for weather prediction was intuitively obvious to a few engineers and scientists in the 1950s (page 130). See RAND publications itemized under chapter thirteen.

Harry Wexler's extensive work in promoting Verner Suomi's experiments to the IGY and in the early days of satellite meteorology (page 131) is obvious from the minutes of the IGY's TPESP, from Wexler's letters to Verner Suomi, from his role as a consultant for Suomi's and Parent's radiation balance experiment (shown by TPESP minutes), and from minutes of the National Research Council's Committee on Meteorological Aspects of Satellites in the immediate post-*Sputnik* days. Wexler died at the age of 50 in 1962.

Sig Fritz's role in the early days (page 131), including his assignment of a broom cupboard for an office, is expounded on in Margaret Courain's Ph.D. thesis, *Technology Reconciliation in the Remote Sensing Era of US Civilian Weather Forecasting*, Rutgers University (1991).

Dave Johnson's participation in both the civilian and defense weather satellite programs is well known among satellite meteorologists (page

131). An unsigned letter to Dave Johnson dated July 29, 1991, which being from Wisconsin must be from either Thomas Haig or Verner Suomi, says, "Delighted to hear that you are about to set the record straight and tell the whole truth about the early met sat days. I'm especially glad that you are the one who is going to do it, because you are really the only one who knew both the civilian and the military programs from the beginning."

The writer puts his finger on the difficulty with writing about the early meteorological satellite days and makes the case for declassification, saying, "I have no clear idea what is still considered to be classified, and I can't imagine why any of the old program history should still be under wraps except perhaps to hide some old CIA–AF feuding that no-one is interested in anyway."

Information about numerical weather prediction (pages 135 and 136) came from my interviews with Joseph Smagorinsky, director of the Geophysical Fluid Dynamics Laboratory in Princeton, New Jersey, from 1970 to 1983. Smagorinsky has been involved in meteorology since his days with the Army Air Corps. He joined the meteorology group of the Institute of Advanced Studies in Princeton in 1950. The group made its first numerical weather predictions on the Electronic Numerical Integrator and Computer (ENIAC);

The beginning of Numerical Weather Prediction, by Joseph Smagorinsky, in *Advances in Geophysics* 25, p. 3 (1983);

John von Neumann, by Norman Macrae, Pantheon Books (1992).

A variety of publications about the Global Atmosphere Research program (pages 132 and 133) are to be found in the library of the National Academy of Sciences. One, published by the International Council of Scientific Unions and the World Meteorological Organization, provides an introduction to the program. It is No. 1 in the GARP Publication series.

Further, less formal, information about the potential role of satellites in the GARP is to be found in a presentation Verner Suomi made to a symposium in October 1969 (the paper doesn't say which symposium, or where). The paper demonstrates Suomi's abilities as a salesman for satellite meteorology.

Chapter thirteen: The Bird's-Eye View

Information about ideas for meteorology satellites in the early 1950s can be found in: RAND's Role in the Evolution of Balloon and Satellite Observation Systems and Related US Space Technology (page 140 and 141), by Merton E. Davies, William R. Harris; and *Inquiry into the Feasibility of Weather reconnaissance from a Satellite Vehicle,* by S.M. Greenfield and W.W. Kellogg. This is an unclassified version of USAF Project RAND Report R-218, April 1951.

An unsigned letter, probably from Thomas Haig or Verner Suomi, talks of the work that the writer and Dave Johnson did toward promoting a single national satellite program in the early 1960s. They were unsuccessful, and both a civilian and military program have since run in parallel. Verner Suomi talks of the duplication he saw (page 147). The writer of the letter to Johnson says of a national program, "I don't think anyone has come close since, and lots of dollars have been wasted as a consequence."

The patent dispute between Hughes and NASA over the spin-scan camera went on for some time. An internal memo from Robert Parent to Verner Suomi of July 8, 1969, outlines the issues and suggests that he and Suomi should put together a chronology in case further action should be taken in future.

Five years later, the dispute was still bubbling along. In a letter dated October 10, 1974, Verner Suomi wrote to Robert Kempf at the Goddard Space Flight Center in Maryland. He described when and how he conceived of the idea for the spin scan camera and what he subsequently did. Suomi asserts that he considers the patent to belong to the U.S. government. The dispute was resolved in NASA's favor.

The report "Space Uses of the Earth's Magnetic Field" (unclassified report) by Ralph B. Hoffman, 1st Lt. USAF, and Thomas O. Haig, Lt. Col. USAF, describes the passive attitude control possible by designing the satellite so that it can take advantage of Earth's magnetic field for attitude control.

Chapter fourteen: Keep it Simple, Suomi

I gathered most biographical details about Verner Suomi from interviews with him and his wife and crosschecked these where possible with written sources and the impressions of those who knew him, which includes nearly everyone in the world of meteorology. I interviewed Dave Johnson, Joseph Smagorinsky, Robert White, Pierre Morel, Thomas Haig, Leo Skille, Bob Sutton, and Bob Ohckers.

The feud between Reid Bryson and Verner Suomi (page 152) is explored by William Broad in the *New York Times* of October 24, 1989.

My favorite piece of correspondence to Wexler, clearly written in response to his efforts to drum up support for Verner Suomi's radiation balance experiment, is from Herbert (Herbie) Riehl, of the University of Chicago (which then had a highly respected meteorology department). Riehl wrote to Wexler on November 28, 1956, from "somewhere over the Rockies" in a plane "with mechanical shakes, hope you have bifocals." He said, "Some hours have gone since our early morning encounter, but they have been enough for my latent astonishment at your remarks over satellites to solidify." Riehl goes on to discuss Earth's net radiation balance. He adds, "I think this is fundamental information for guiding meteorological research on long (and very long) period changes."

Wexler presented Suomi's idea for a radiation balance experiment to the Technical Panel on the Earth Satellite Program (page 154) on the second day of the sixth meeting of the TPESP on June 8, 1956. James Van Allen, with his credentials as the former chair of the Upper Atmosphere Research Panel, headed a Working Group on Internal Instrumentation formed by the TPESP at its third meeting, on January 28. The UARP had received many suggestions for satellite instrumentation following President Eisenhower's announcement of July 29, 1955. One of these experiments was that of Bill Stroud, from the Signal Corps of Engineers. Van Allen pointed out that Stroud's experiment had already been approved, and that while not as broad as Wexler's proposal, it was simpler.

Wexler obtained the backing of the IGY's Technical Panel on Meteorology, of which Wexler was chair, for both Stroud's and Suomi's experiments at the TPM's eighth meeting, on October 9, 1956.

Suomi and Parent received their formal go-ahead to produce their satellite for a Vanguard launch on December 31, 1957, from J.G. Reid, secretary to the TPESP.

The account of the Juno II explosion (page 161) comes from *Wisconsin State Journal*, July 17, 1959.

A draft of the IGY terminal report of Suomi's radiation balance experiment prepared July 27, 1961, by Stanley Ruttenberg, head of the IGY' program office, gives details of the instrumentation in operation (page 162). It says, "A huge amount of data is accumulating from this experiment ... only a start has yet been made on reducing this data and analyzing it." Daytime data was less useful than nighttime data, noted the report, because of interference from the ionosphere. It says, "Despite the necessary shortcomings of the data there does seem to be a clear indication that large scale outward radiation flux patterns exist and that these patterns are related to the large scale features of the weather." The report concludes, "The experience being gained from this experiment will be an important factor in designing future meteorological satellite experiments."

Chapter fifteen: Storm Patrol

Details scattered through chapter 15 come from the following: Interviews with Verner Suomi, Bob Ohckers, Bob Sutton, and Leo Skille.

"Initial Technical proposal for a 'Storm Patrol' Meteorological Experiment on an ATS Spacecraft" by V.E. Suomi and R.J. Parent, September 28, 1964;

"Spin Scan Camera System for a Synchronous Satellite" prepared for NASA by V.E. Suomi and R.J. Parent—July 1965; and "The Spin Scan Camera System: Geostationary Meteorological Satellite Workhorse for a Decade." Optical Engineering 17 (1), January–February 1978.

Books of Value for the Meteorology Section

Planet Earth: The View from Space, by D. James Baker (Harvard University Press, 1990).

Watching the World's Weather, by W. J. Burroughs (Cambridge University Press, 1991).

Weather Cycles: Real or Imaginary, by W. J. Burroughs (Cambridge University Press, 1992).

Communications section

There was so much material for this section that the only way to make sense of it was to put it all together into one big pot, to arrange it chronologically, and to construct a series of calendars for the years in which I was interested. I also put the dates of major importance for world events and other developments in the space program on the same calendars. This gave me a good feel for what was happening when, and highlighted some nice ironies between the Telstar and Syncom programs that I would otherwise have missed.

Primary source material came from AT&T, the Hughes Aircraft Company, John Rubel, Bob Roney, the NAS, NASA, and the American Heritage Center.

Interviewees were John Pierce, Harold Rosen, Tom Hudspeth, Bob Roney, Robert Davis.

AT&T's highly professional archive yielded masses of information about Telstar and some, though to a lesser extent, about Echo.

The HAC archives give a good sense of the work that Harold Rosen and Don Williams et al. did, as well as demonstrating the company's internal wrangles and its lobbying of NASA and the DoD.

Bob Roney had personal papers that supplemented the more extensive records from the HAC archive.

John Rubel's papers cover the ground from the perspective of the Office of Defense Research & Engineering. The view from this office is not always the same as that from other departments of the DoD.

NASA's records (thanks once again to David Whalen for these) gave, not surprisingly, the agency's side of the story.

I had more material from AT&T and the HAC and John Rubel than from NASA, so the story unfolds primarily from their perspective, though I have tried to synthesize different policy and technical viewpoints.

Chapter sixteen: The Players

TAT-1 capacity (page 171) from *Signals, The Science of Telecommunication,* by John Pierce and Michael Noll. Scientific American Library (1990).

The Space Station (page 172), Its Radio Applications, by Arthur C. Clarke, 25 May 1945 (typed manuscript).

"Extra-Terrestrial Relays, Can Rocket Stations Give World-Wide Radio Coverage," by Arthur C. Clarke, published in *Wireless World,* October 1945 (page 172) (From John Pierce).

"Orbital Radio Relays," by J. R. Pierce, *Jet Propulsion,* April 1955 (pages 172–173).

Pierce's views of Rudi Kompfner, medium-altitude satellites, and Harold Rosen (pages 174–175) from interviews with Pierce, his oral history at Caltech, memos, and his autobiography (see notes for chapter 17).

All of the following documents provided details on which the discussion of policy on pages 176–179 are based.

Management of Advent (March 1961): including TAB F—Memorandum from the Acting Secretary of Defense of the Army and the Secretary of the Air Force, subject, Program Management of Advent (15 September, 1960); and TAB G—Recommended Action: Memo for Signature of Secretary of Defense.

The problems in the Advent program were spelled out by, among others, Harold Brown, DDR&E, on May 22, 1962, in a memo for the secretary of defense, Robert McNamara. In an appendix, Brown states categorically that the contractor's management of Advent (General Electric) was poor. He also wrote, "During 1961 and until February 28, 1962, USAAMA (Army Advent Management Agency) decided to gamble on the contractors to realize first a March and then a June firing date. The rate of expenditures rose to twice the amount allowable on the basis of available funds and the whole project went out of balance; training, ground equipment and operational control rooms were fully engineered before the spaceborne equipment was out of the breadboard stage; schedules were manipulated to a point where the completion date of the first flight article was set ahead of the engineering test model" (John Rubel's papers).

Memorandum for John H. Rubel, Deputy Director of Defense Research & Engineering, from Ralph L. Clark, Assistant Director for Communications, dated January 17, 1960. A number of policy documents on communication satellites are attached. These are: Briefing paper for Mr. Gates, subject: Cabinet paper CP60-112—Communication Satellite Development, December 19, 1960; Draft policy presented by ODRE to the unmanned spacecraft panel of the AACB; Summary of Cabinet Paper CP60-112/1, December 23, 1960; Briefing paper by Clark and Nadler setting out their concerns with the Cabinet Paper; draft of a policy statement prepared by the service secretaries summarizing the DoD's role and interest in communication satellites (John Rubel's papers).

Memorandum for Members of the Unmanned Spacecraft Panel: Statement of NASA Program Philosophy on Communication Satellites. November 21, 1960 (HAC archives).

The NASA Communication Satellite Program, February 9, 1961 (John Rubel's papers).

"United States Policy Toward Satellite-Based Telecommunications," circulated among a small group by John Rubel in April 1961 (John Rubel's papers).

Informal Notes of the Interim Steering Committee for Satellite-Based Telecommunications Policy, second meeting, held in the Pentagon, 11 May, 1961. Appendix B was: "DoD Position on what Technical Characteristics and Capabilities DoD Desires from a Commercially Operable Satellite-Based Telecommunication System". Appendix C: Industry-Department of Defense Cooperation in Satellite-Based Telecommunications (John Rubel's papers).

A memorandum dated September 6, 1960, records a meeting at AT&T's headquarters on that day. The meeting discussed a request from NASA for information about Bell System's plans for satellite communication and research. Memo gives AT&T's policy views. The views were laid out in a letter of September 9 to T. Keith Glennan, which said that satellites should be operated by commercial companies, not government, and that enough information existed from Echo 1 for Bell to want to proceed immediately to work on active repeaters. (Box 85080203, AT&T archives).

AT&T was clearly fighting for a comprehensive role in satellite communication, as numerous documents in AT&T's and NASA's records show. For example; on March 10, 1961, Jim Fisk, head of BTL, sent a letter to Richard S. Morse, assistant secretary of the Army, expanding on an informal proposal sent by Bell to Morse on March 3. Even though the letter refers to Bell's ideas for an experimental, not operational, program, the proposal's completeness, with all the control that it would have ceded to AT&T, might reasonably have raised concerns in government over the extent of the control that the monopolistic AT&T would yield over something as vital as international communications.

Fisk wrote, "Bell Systems' interest is simply stated: communication satellites promise a natural extension of the present microwave common-carrier networks and a natural supplement to present overseas radio and cable circuits."

Specifically, Fisk proposed that Bell should design, construct, and pay for the fixed ground stations in the U.S.; arrange for foreign ground stations with overseas common carrier partners; design, construct and pay for repeaters, providing frequencies for specific military uses as well as common carrier uses; accommodate the experimental requirements of the other common carriers on terms mutually agreeable; provide systems engineering assistance to the Department of Defense from the development of transportable or mobile ground terminals; provide systems engineering assistance to the Department of Defense to adapt the low-orbit satellite repeaters into synchronous orbit repeaters when the orbiting, orientation, stabilization and station keeping problems of that satellite are solved; cooperate with the Department of Defense in the initial launch operation, and share the costs of launching, as may be agreed; work with other Department of Defense contractors on portions of the program of primary military interest to insure efficient planning and to insure system compatibility.

The tensions between NASA and AT&T at both policy and technical levels are also well documented. A letter from Fred Kappel, the president of AT&T, to James Webb, the administrator of NASA, written on April 5, 1961, says, "It has come to my attention that an article that *The Wall Street Journal* carries ... that NASA has yet to receive any firm proposal from any company." Kappel goes on to write, "In view of the events which have taken place during the past few months, this statement ... is of deep concern to me. The specific events to which I refer are as follow." Over four pages, Kappel itemizes approaches made by AT&T to NASA (George Washington University, passed to me by David Whalen).

The Department of Justice's concerns about the antitrust implications of AT&T's plans for an operational communication satellite system are mentioned in various places. One source is a memo for Alan Shapley from James Webb, NASA administrator, dated August 12, 1966. In this memo Webb also makes the point that the RCA proposal (Relay) was clearly the best for the "experimental and research requirements of NASA, although not necessarily the best for the first step toward an operational communication satellite as desired by AT&T" (NASA History Office).

A memorandum for the record by Robert Nunn of December 23, 1960, describes a meeting between himself; John Johnson, NASA's special council; and Attorney General Roberts. They were discussing a paper on communication policy to be submitted to the White House. December 1960 was, of course, the eve of the Kennedy administration. The policy question at stake was private or public promotion of communication satellites. All acknowledged that AT&T might be the only company capable of owning and operating an operational system of communications satellites. Roberts said, "Whatever we do, we cannot act as though NASA is putting AT&T into a preemptive position ..." And he said, "... we cannot assume ... that when all is said and done AT&T will emerge owning and operating a system. . . ." Nunn explained, " . . . the feeling in the White House apparently favors taking a position on principle which the succeeding administration will be obliged to overturn if it does not concur. Rogers said that this was unrealistic. Their aim was to find a way of supporting the White House's stance in favor of the private sector without "seeking to nail down the conclusion concerning what the government will or will not do in future."

A memorandum for the special assistant to the administrator at this time spells out AT&T's dominance of international communication. The voice segment (cable and radio) was operated exclusively by AT&T. Some nineteen companies competed for telegraph traffic (NASA History Office).

A number of policy documents and internal ODR&E memos point out that the solar minimum of 1964 to 1966 would reduce radio communication circuits by up to two-thirds. These include a letter from Jerome Wiesner, who wrote on July 7, 1961, to Robert McNamara urging him to give his personal attention to the strengthening of the Defense Communications Agency.

The cooperation between NASA and the DoD with respect to which organization would develop which satellites (at least at the highest levels) is apparent in a letter from James Webb to Robert McNamara, secretary of defense. On June 1, 1961, Webb wrote, "I also wish to take this opportunity again to make clear my firm intent that you are kept informed of activities concerning communication satellites and that your views and interests are kept in mind at all times." At lower levels, and even within the separate organizations, relationships were not always so open. A memo from James Webb to Hugh Dryden, dated June 16, 1961, says, "I think it important that we not ever indicate that some of our military friends, particularly those down the line in the services, may not have had full access to all the information, documents, and so forth relating to the kind of decisions that Mr. McNamara, Mr. Gilpatric, and I have made on the big program" (David Whalen from NASA History Office).
The issue being discussed at this time was NASA's involvement in synchronous altitude, or 24-hour, communication satellites. The succeeding agreements that NASA and the DoD had, first that NASA should develop only passive satellites and then that it develop only medium-altitude satellites was explained to me by John Rubel.

Letter from James Webb, NASA administrator, to the director of the Bureau of the Budget, dated March 13, 1961. On communication satellites, Webb wrote, "This proposal specifically contemplates that this Administration should reverse the Eisenhower policy under which $10 million of the NASA active communication satellite program would be financed by private industry. In my view this would not be in the public interest at this stage in a highly experimental research and development program" (David Whalen, NASA History Office).

A memorandum, dated April 28, 1961, for John A. Johnson, NASA general council, from James Webb, on the subject of a letter to Fred Kappel, president of AT&T, gives Webb's view that government should not reach a hasty decision about its policy on experimental programs, such as communication satellites, without thoroughly considering all possible interest groups (David Whalen from the NASA History Office).

Memorandum for the associate administrator, dated May 16, 1961, concerning expansion of the active communication satellite program, by Robert Nunn. The memo discusses the policy issues and the public versus private development of operational communication satellites.

Summary of the Comsat Bill:
Comsat stock was issued on June 2, 1964. There were ten million shares at $20 per share. The net proceeds to Comsat were $196 million. Half the capital was raised from individuals and half from 150 communications companies.

Chapter seventeen: Of Moons and Balloons

Personal details about John Pierce (mainly pages 180–182) come from my own interviews with him.

Other details are from the next four references: oral history taken by Harriet Lyle for the archive of the California Institute of Technology in 1979.

The Beginnings of Satellite Communications, by J. R. Pierce, with a preface by Arthur C. Clarke (San Francisco Press, 1968) (copy from John Pierce);

My Career as an Engineer, An Autobiographical Sketch, by John R. Pierce, the University of Tokyo (1988); and

"Orbital Radio Relays," by J. R. Pierce, *Jet Propulsion,* April 1955 (page 183).

In Spring 1958, Pierce and Rudi Kompfner read about William O'Sullivan's ideas for a balloon-like satellite to measure air density and realized that by bouncing microwaves off its surface they could test many technical aspects of satellite communication (page 184). Interview with John Pierce.

Brief notes describing an ARPA satellite conference on July 13–14, 1958, and John Pierce's involvement with William Pickering in developing the Echo project (page 184). The notes are in the John Pierce collection (#8309 Box 5, folder-leading to Echo) at The American Heritage Center at the University of Wyoming.

At the sixth meeting of the TPESP on June 7 and 8, 1955, William J. O'Sullivan, from the National Advisory Committee for Aeronautics (NACA—a forerunner of NASA), presented his ideas for a giant aluminized balloon that could be observed from Earth, allowing information to be deduced about the density of the upper atmosphere (page 184). Richard Porter told him the idea was interesting, and that the proposal should be sent to the working group on internal instrumentation (NAS archives).

At the seventh meeting of the TPESP, on September 5, 1956, O'Sullivan told the panel that NACA was prepared to build its air drag satellite experiment without the backing of the IGY (NAS archives).

There is a May 13, 1958, memo from Rudy Kompfner to E. I. Green, including John Pierce's memo "Transoceanic Communication by Means of a Satellite."

May 26, 1958: a memo by Brockway McMillan, "A Preliminary Engineering Study of Satellite Reflected Radio Systems." The study is based on Pierce's ideas. McMillan favors twenty satellites at medium altitudes to give continuous service and envisages that there would be a market for a new transatlantic service between 1970 and 1975. He writes, "It is concluded here also . . . there probably exists a potentially much larger market. . . ."

There is a July 25, 1958 memo from E. I. Green to Mervin Kelly evaluating Pierce's proposals concludes, "In summary, the proposed system would require intensive R&D on a host of problems . . . considering other demands of Bells Systems . . . it would be my recommendation that we do not attempt to undertake satellite communications as a Bell System development." Presumably, it was this memo that led to his "cease and desist order" (page 185).

A letter from John Pierce to Chaplin Cutler of October 17, 1958, gives Pierce's views of the meeting of the Advanced Research Project Agency he had attended on October 15 and 16 (page 185). Pierce was aware from the beginning of the ideas being considered by the military for communication satellites. At the ARPA meeting he was acting as a consultant to what was an ad hoc panel on 24-hour satellites. ARPA's views would change considerably. Pierce reported the view then: "It is possible that a spinning satellite with non-directive antennas will be launched early in 1960 and a satellite with an attitude stabilized platform in 1962" (Box 840902, AT&T archives).

A memorandum for the Record from John Pierce, Rudy Kompfner, and Chaplin Cutler on Research Toward Satellite Communication, Research toward Satellite Communication. Dated January 6, 1959, the memo describes a research program directed in general at acquiring the basic knowledge for satellite communication by any means and specifically at aspects of passive Echo-type satellites. A fuller version of the research memo was written on January 9, 1959 (page 185) (AT&T archives).

In a letter to William McRae, vice president of AT&T, of January 7, 1959, Pierce outlines his proposed research program for satellite communication (AT&T archives).

A memo of 16 June 1959 describes a meeting between BTL, the Jet Propulsion Laboratory, and the Naval Research Laboratory, during which William O'Sullivan provided some technical details on the aluminized balloon and the Langley Research Center's tests (page 186) (AT&T archives).

Information scattered through the chapter came from: Monthly Project Echo reports starting October 23 1959 (AT&T archives);

Project Echo, Monthly report No. 3, December 1959. Report on the first moon bounce test;

Rudi Kompfner's correspondence and memos (AT&T archives 59 04 01); and from

Film reels in the AT&T archives:
1. Project Echo 1. An NBC news special sponsored by BTL 409-0213
2. The Big Bounce, BTL film 399-03727.

Chapter eighteen: Telstar

Documents drawn on for the launch of Telstar:

Satellite ground tracking station, Andover, Maine, Engineering notes: Telstar July 9, 10, and 11, 1962. The document gives details of the countdown (page 188), for example, loss of calibration by the ground tracker at 1220 UT, power supply trouble at 2317 in the upper room of the communication antenna, etc. . . . (box 85080302 - AT&T archives).

Memorandum for the Record from John Pierce, Rudy Kompfner, and Chaplin Cutler on Research Toward Satellite Communication, and Research toward Satellite Communication (page 189). Both are dated January 6, 1959, and deal with a research program directed in general at acquiring the basic knowledge for satellite communication by any means and specifically at aspects of passive Echo-type satellites. A fuller version of the research memo was written on January 9, 1959 (AT&T archives).

In this chapter references to what NASA officials said or did (pages 191 to 198) comes from documents in the NASA History Office or George Washington University. These were shared with me by David Whalen.

They include:

Memorandum for the Record, October 31, 1960, by Robert G. Nunn, special assistant to the administrator. This summarizes a meeting between NASA officials and James Fisk, president of Bell Telephone laboratories, and George Best, vice president of AT&T. The purpose was to discuss Bell's plans for "Transoceanic Communication via satellite." It opened with T. Keith Glennan, NASA's administrator, saying that Bell had not considered the "facts of life" with respect to vehicle availability. The meeting discussed policy issues in some depth, including finance and whether or why public money should be spent on communication satellites.

Memorandum for program directors, February 24, 1961. Subject: Guidelines for preparation of preliminary Fiscal Year 1963 budget. On the subject of communication satellites, it said to assume no funding of operational systems; adequate provision should be made for back-up vehicles; and no development of passive communication systems.

Minutes of the administrator's staff meeting: November 30, 1960; December 1, 1960; December 8, 1960; January 18, 1961; January 26, 1961; February 2, 1961; March 2, 1961; May 25, 1961; June 1, 1961; June 12, 1961; June 15, 1961; June 22, 1961; June 29, 1961.

Technical details about Telstar and the attitudes and opinions of the Bell engineers were gleaned from the following:

"Project Telstar, Preliminary report Telstar 1 July–September 1962." (AT&T archives).

Telstar—*The Management Story,* by A. C. Dickieson (unpublished manuscript, July 1970). Dickieson was the project manager for Telstar (AT&T archives).

Extracts from a manuscript by D. F. Hoth. Chapter on Telstar Planning: January–May 1960 (AT&T Archives 84-0902).

Each quotes extensively from memos that the writers had access to.

The discussions of technology in the chapter come from a mixture of sources, including documents in the Hughes Aircraft Company's archives.

Helpful textbooks include:

Satellite Communication Systems Engineering, by Wilbur L. Pritchard, Henri G. Suyerhoud, and Robert A. Nelson (Prentice Hall, 1993).

The Communication Satellite, by Mark Williamson (Adam Hilger, 1990).

Chapter nineteen: The Whippersnapper

What Pat Hyland thought about Syncom's early development is found in an extensive video interview recorded on December 14, 1989 (page 199–201). Copy available from HAC.

HAC's early views on the commercial opportunities of space come from Bob Roney (page 201).

Frank Carver's request that Harold Rosen look for new business ventures (page 201) was remembered by Harold Rosen and Bob Roney in their interviews with me.

The account of Rosen's actions and discussion with Williams come from my interview with Rosen (page 202).

The account of Roney's recruitment of Williams comes from my interview with Roney (page 202).

The account of Rosen's efforts to tempt Williams back to Hughes comes from my interview with Rosen (page 202).

The technology in pages 203 and 204 is my distillation of the technical information in a number of memos, proposals, textbooks, and interviews.

Rosen's attraction to Southern Californian beach parties is his own recollection in an interview with me (page 204).

Sydney Metzger's comment (page 204) was made during an interview dated December 5, 1985, when he said, "When we (RCA) heard of Syncom we could have kicked ourselves for not thinking of a spinner at synchronous altitudes since RCA had the very early spinner experience." Metzger, who worked for RCA, joined Comsat in June 1963 as the manager of engineering (HAC archives 1993-50 Box 1).

Comments on the TWT for HAC's 24-hour satellite made in interviews with Tom Hudspeth, Rosen and Roney (pages 204–205).

The date that Leroy Tillotson sent his proposal for a medium-altitude satellite to Bell's research department (page 205) is given in A. C. Dickieson's book (see notes for chapter 18).

Carver's and Puckett's immediate views of Rosen's proposal were given in Rosen's interview with me (page 205).

A memo from A. S. Jerrems to F. R. Carver on September 17, 1959, reminds Carver of a meeting planned for September 23 to work up a presentation on communication satellites for Allen Puckett (page 205) (HAC archives).

Rosen and Williams first describe their satellite in "Commercial Communication Satellite," October 1959, by H. A. Rosen and D. D. Williams (page 205), and in Preliminary design analysis of communication satellite, October 1959. This paper reviews the torque box design that Harold Rosen and Don Williams put forward for a 24-hour satellite in September (From HAC archives).

Sam Lutz examined the Rosen Williams idea. His evaluation appears in a memo from S. G. Lutz to A. V. Haeff, October 1, 1959, Evaluation of H. A. Rosen's commercial satellite communication proposal (From Bob Roney) (page 207).

A memo from A. S. Jerrems of October 9, 1959, confirms the establishment of a two-week-long intensive study of the Rosen proposal (page 207).

A memo from S. G. Lutz to A. V. Haeff on October 13, 1959. Subject: Economic aspects of satellite communication gives Lutz's opinions (page 207).

Memo from J. H. Striebel to A. V. Haeff of October 22, 1959. Subject: market study for a worldwide communication system for commercial use shows more of the thinking at HAC (page 207).

Lutz's second evaluation of the Rosen Williams proposal appears in a memo from S. G. Lutz to A. V. Haeff of October 22, 1959 (page 207). Subject: commercial satellite communication project; preliminary report on study task force.

A memo from L. A. Hyland to A. E. Haeff and C. G. Murphy of October 26, 1959. Subject: communication satellite orders an immediate and comprehensive study should be made of patentable potentialities and NASA's position should be ascertained (page 207). A number of subsequent memos show that Hyland's instructions were carried out. Invention disclosure was November 2, 1959.

A memo from D. D. Williams to D. F. Doody on November 23, 1959 described Williams's talks on November 5 with Homer Stewart, then at NASA, during which Williams emphasized that Hughes wished to maintain its proprietary and patent rights and the company's desire that the project should be undertaken as a commercial venture. The two also discussed technical issues (page 208).

An interesting aside given the later legal action over patents between HAC and NASA is found in a memo from David Doody to Noel Hammond saying that should a 30-day analysis then being undertaken by the company show the 24-hour satellite to be feasible, Hughes would attempt to win a contract from NASA and would proceed with filing a patent application prior to contracting with NASA. He said further that the company would not yet enter the communication field or approach communication companies with the proposal. He further wrote, "We will take our chances on retaining title to the inventions that have been made to date, but should NASA insist on taking title as a result of supporting the development, the company will go along with NASA since it does not intend to use resulting patents primarily for the purpose of enhancing its patent holdings." This view is at odds with the decades-long battle that Hughes fought with NASA.

In September 1959, a barrage of technical memos begins covering topics such as dynamic aspects of communication project, feasibility investigation of payload electronics. The technical memos mushroom during the following years.

Despite Hyland's decision not to commit funds to the 24-hour satellite (page 208) Rosen and Williams write *"Commercial Communication Satellite", January 1960,* by H. A. Rosen and D. D. Williams. By now the 24-hour satellite has the familiar cylindrical shape.

A memo from Robert Roney to A. E. Puckett of 27 January 1960. Subject: communication satellite review analysis (From Bob Roney) describes yet another review of the Rosen/Williams idea (page 208).

On March 23, 1960, Williams wrote to Hyland, saying that he was pleased by Hyland's decision to fund the commercial communication satellite. He wrote, "It is my understanding that the program will ultimately be financed by sources of capital external to the company. As one of the inventors of the system, I would like to invest in it myself if possible. I enclose a cashier's cheque for $10,000. While I realize that this amount will not go very far, I think it can be multiplied by 100 if the company is willing to permit investment by its employees." This was after Rosen, Hudspeth, and he decided to find some of their own money for the project (page 209).

A memo from Allen Puckett to D. F. Doody dated March 7, 1960 details the requests by Williams, Rosen, and Hudspeth to be released from their usual patent agreements should Hughes not go ahead with the development of a communication satellite. Puckett states that their request is reasonable (page 210).

Details of Rosen's attempts to raise money from various sources are from my interviews with Rosen.

Chapter twenty: Syncom

Chapter 20 is a distillation of information from the following documents, presented in chronological order, though extracts from one document or an appendix may have been useful for explaining some other part of the unfolding events.

Interviews with Rosen and Roney provided detail that do not appear in the written record. I have included it where it seemed to make sense. For example, it was Rosen who told me that T. Keith Glennan first told Puckett that he was "talking through his hat" when Puckett presented the company's idea for a 24-hour satellite (page 212). Though Glennan was interested in what Puckett said, Glennan knew that HAC then knew nothing about satellites and that the 24-hour satellite idea was far from conservative. Glennan might well have said what Rosen recalls he said.

April 26, 1960 Rosen completes evaluation of life and reliability of the proposed satellite, focusing on electronics and TWT design. A handwritten note from Puckett says, "This looks very promising. Thanks."

May 1960: A hefty, mathematical document from Williams entitled "Dynamic Analysis and Design of the Synchronous Communication Satellite."

A memo from J. W. Ludwig to C. G. Murphy and A. E. Puckett of May 2, 1960, discusses a meeting with E. G. Witting, of the Army, and a representative of the Office of Defense Research and Engineering (Mr. Evans). Hughes learned that the Army was already considering other 24-hour satellite proposals and that Herb York, DDR&E, was "intensely interested in the Hughes program."

On May 19, J. W. Ludwig sent a memo to A. E. Puckett about a forthcoming request for proposals (July 1, 1960) from the Army for the Advent 24-hour communication satellite.

June 1960: Synchronous communication satellite, proposed NASA experimental program by the HAC Airborne Systems Group. Proposal included details of Jarvis Island where HAC was at that time proposing it should build a lunch site.

A memo from Lutz for file copied to Rosen on June 3, 1960, summarizes in detail presentations by Pierce, Jakes, and Tillotson from Bell Telephone Laboratories at a conference at the end of May.

Letter from Douglas Lord, technical assistant to the Space Science Panel, thanking Allen Puckett for the Hughes presentation to the President's Science Advisory Committee.

Harold Rosen told me the anecdote of how the meeting came to be set up (page 212).

July 26, 1960, Puckett confirms a meeting requested by Abe Silverstein for Hughes to present its satellite proposal to Keith Glennan.

Technical memos in the meantime show Williams's preoccupation with dynamics studies and his involvement with NASA's Langley field center.

Memo from John Richardson to C. G. Murphy of August 12, 1960, talks of GT&E's evaluation of the Hughes satellite. Richardson believed that the GT&E reaction was quite good. GT&E asked Hughes for justification of the life expected for the TWT; how the telemetry system could be protected from disruptive tampering and how the Hughes assertion that ten launches would be needed for one successful satellite could be reconciled with the company's plans for three launches.

An internal memo from Ralph B. Reade to Roy Wendahl raised concern about internal conflicts in the field of satellite communication and the company's fragmented approach to military customers.

September 14, 1960: preliminary cost estimates of a commercial communication satellite. Total: $15.75 million, including $4 million for the Jarvis Island launch site.

A letter from John Rubel to A. S. Jerrems dated September 22, 1960, refers to a visit he had made a few weeks earlier to Hughes.

On October 25, 1960, Rosen wrote back to Witting refuting all the specific criticisms that Witting made of the HAC proposals. Rosen concluded, "In our opinion, the Hughes proposal, if implemented, would achieve all the objectives of the present program, but at an earlier date and lower cost."

Memos in September, October, and November show that HAC held meetings with GT&E, The Rand Corporation, Bell Telephone Laboratories (November 2, 1960), ITT, and British Telecommunications.

A Letter from Allen Puckett to Lee DuBridges, president of Caltech, written on November 18, 1960, refers to their discussions about communication satellites at the Cosmos Club.

Memo from Allen Puckett to J. W. Ludwig of December 14, 1960, concerning NASA's forthcoming RFP for a medium-altitude active satellite.

Memo from Bob Roney to Allen Puckett on December 23, 1960, concerning the need for budget decisions for the 24-hour communication satellite program for 1961.

In a memo of January 6, 1961, Puckett informed Rosen, Williams, Hudspeth, and a few others that Lutz would once again be evaluating their communication satellite proposal.

An agenda of an Institute of Defense Analysis meeting shows Rosen scheduled to brief the group between 10:30 and 11:00 on January 10, 1961. Rosen's write-up of the meeting on January 12 was directed at Allen Puckett. I noted that the panel was critical of Advent and that his description of the Hughes 24-hour satellite elicited "generally favorable comment."

In a memo to C. G. Murphy and F. P. Adler, dated January 13, 1961, Sam Lutz demonstrated a less enthusiastic view of the 24-hour satellite, advocating extensive additional work and comparisons with passive and medium-altitude active satellites.

Memo from S. G. Lutz to Allen Puckett and A. V. Haeff of February 10, 1961. Subject: review of satellite communications. Lutz wrote a negative and critical report of the 24-hour satellite.

Telegram from Rosen to Rubel of February 28, 1961, apparently responding to remarks by Bill Baker, of Bell Telephone Laboratories.

March 1961: HAC report, "Stationary Satellite, Island Operation Phase."

Memo from Allen Puckett to F. P. Adler of April 20, 1961. Subject: a conversation Puckett had had with John Rubel and Rubel's suggestions regarding the actions that Hughes should take (HAC archives: Commercial Communication Satellites 1990–05, Box 6 DoD Communication Satellite).

Memo from C. Gordon Murphy to J. W. Ludwig of April 27, 1961. Subject: A conversation Murphy had had with Rubel.

Letter dated May 8, 1961, from Allen Puckett to John Rubel. Subject: A Program for Interim Satellite Communication.

Memo dated May 11, 1961, from Williams to Hyland, outlining the mistakes he thought the company had made with respect to developing the 24-hour communication satellite.

Telegram dated May 18, 1961, from Jack Philips to R. E. Wendahl, telling him that Hughes had just been informed that they were unsuccessful in their bid for Project Relay.

Memorandum for the Associate Administrator from Robert Nunn and Leonard Jaffe to Robert Seamans, dated June 6, 1961. The memo states, "It is recommended that NASA immediately embark on a project leading to tests of a simple, light weight communication satellite at 24-hour orbital altitudes. . . ." This memo formalized a situation that Seamans and Rubel had worked to bring about.

Memo from A. S. Jerrems to Allen Puckett dated June 19, 1961. Jerrems wrote, "Here is another bulletin on the Rubel situation. In a telephone conversation this weekend, Rubel advised me that he spent all day Saturday (June 17) in a meeting with NASA to discuss communication satellite plans. He was inscrutable about the detailed content of the meeting, but he made a statement to the effect that in his opinion, HAC prospects for getting a synchronous satellite funded were better now than they had ever been."

A letter of June 23, 1961, from Roswell Gilpatric, deputy secretary of defense, to James Webb. He writes, "Mr. Rubel has told me about the plans that he and Dr. Seamans are formulating for an interim synchronous satellite communication experiment with potential for an interim opera-

tional capability within the next 12 to 24 months. . . . I regard the proposed program as complementary to those [Rely and rebound (Echo-type
satellite)] and the Advent project. . . . You have my assurance of support in
the event that, in your judgement, their proposals should be adopted."
With this letter, Gilpatric set aside the informal agreement that NASA
would not develop synchronous-altitude satellites.

Memo from A. S. Jerrems to J. H. Richardson, Allen Puckett, and R. E.
Wendahl of July 20, 1961, on the possibility of an early synchronous-orbit
experiment. HAC still did not know that both NASA and the DoD were
proposing a sole-source contract. After dinner and talks with Rubel, Jerrems wrote, "Although the proposed five week study program for
Hughes, which Rubel and Seamans described to us at the end of June, has
not yet been kicked off as we hoped it would be, the planning for the
Special Program is not quiescent. There have been a continuing series of
meetings between NASA and the DoD to iron out the definition of the
ground roles for HAC."

On August 9, 1961, Alton Jones, project manager at the Goddard Space
Flight Center and James McNaul, acting project manager for the U.S.
Army Advent Management Agency, signed a contract to be jointly pursued by NASA and the DoD for the preliminary project development
plan for a lightweight, spin-stabilized communication satellite.

On August 12, 1961, Maj. Gen. G. W. Power, director of developments in
the Office of the Chief of Research and Development, wrote to Rosen
rejecting his ideas for a lightweight, spin-stabilized communication satellite.

Bell Systems was well aware of the promise of communication satellites at
synchronous altitudes and also aware of the station-keeping difficulties. In
a paper written in November 1962, K. G. McKay wrote an internal paper
on the pros and cons of synchronous satellites. He wrote, "The synchronous satellite must be placed at a specific point in space with exactly the
right velocity and kept there for the life of the satellite. It is a bold concept and I am confident that some day it will be achieved" (box 840902 -
AT&T archives).

Rejections of the Hughes Proposal

From Maj. Gen. Marcus Cooper, Air Research and Development Command, to Allen Puckett on 3 January 1961: "Since your 3 November visit here ... analyzed in detail the Hughes Aircraft Company proposal for a synchronous altitude active communication satellite. In general their findings reveal that the proposal is technically marginal in several respects, and tends to be overly optimistic...." Cooper goes on to say that in view of the results expected from Advent, he did not think that "we should proceed further with the Hughes proposal at this time."

From E. G. Witting, deputy director of research and development, to Mr. J. Bartz, assistant manager, for contracts, writing on October 11, 1960, in response to a letter from HAC of April 22, 1960. Witting writes, "Your proposal has been thoroughly evaluated by the Army and it has been determined that the project would not meet the present requirements of the Army for intercontinental communications."

Documents that contributed to the general framework of the chapter:

"Syncom (Interim Communication Satellite) Chronology," possibly prepared in the spring or summer of 1963, giving dates of John Rubel's involvement with Syncom between 10 April 1961 and 6 February 1963.

Advent chronology from 14 April 1960 to December 1961 (John Rubel's papers).

Policy statement for exploitation of HAC communication satellite, undated and unsigned.

Preliminary history of the Origins of Syncom, by Edward W. Morse (NASA Historical Note No. 44) September 1, 1964. Some aspects of this report, about agreements between NASA and the DoD for instance, confirm details in earlier chapters in this section (John Rubel's papers).

Although John Rubel was interested in Syncom, others in the Office of Defense Research and Engineering were less keen. Dr. Eugene Fubini, for example, was not (Author's interview with John Rubel). In a memo dated March 26, 1962, Fubini was still arguing strongly in favor of Advent (John Rubel's papers).

R. H. Edwards to D. D. Williams, 19 January 62, "Separation of syncom payload from the third stage" (HAC archives 1987-44 box 1).

"Torques and Attitude Sensing in Spin-Stabilized Synchronous Satellites," by D. D. Williams, American Astronautical Symposium, Goddard Memorial Symposium, March 16–17, 1962.

Post Syncom decision:

Interest in Syncom grew once it had become an official project. An internal HAC memo date 9 May 1962 from C. Gordon Murphy to R. E. Wendahl discussed a visit by the commanding general of the U.S. Army Advent Management Agency, who was interested in HAC's ability to provide a replacement for the Advent Spacecraft.

By June 18, Robert Seamans was writing to John Rubel about NASA's plans for a follow-on Syncom program—a five-hundred-pound spacecraft that would permit the "incorporation of 4 independent wideband transponders, redundant control systems and sufficient on-board auxiliary power to operate the system continuously." Such a satellite, as a memo from Robert S. McNamara, dated May 23, 1962, shows, would provide a suitable alternative to Advent (John Rubel's papers).

Memo from John Rubel, deputy DDR&E, to the assistant secretary of the Army, January 25, 1962. Subject: DoD support of NASA–Syncom communication satellite test (John Rubel's papers).

Documents for general background to the communication section

"Telephones, People and Machines," by J. R. Pierce, *Atlantic Monthly,* December 1957.

"Transoceanic Communication by Means of Satellite," by J. R. Pierce and R. Kompfner, *Proceedings of the IRE,* March 1959 (David Whalen, from George Washington University).

"Satellites for World Communication," report of the Committee on Science and Astronautics, U.S. House of Representatives, May 7, 1959.

Project Summary: Project Courier Delayed Repeater Communications Satellite. November 17, 1960 (John Rubel's papers).

Memorandum for the President, presented to Cabinet December 20, 1960 (David Whalen, from George Washington University).

A Chronology of Missile and Astronautic Events, Report of the Committee on Science and Astronautics, U.S. House of representatives, March 8, 1961.

Special Message to the Congress on Urgent National Needs, delivered to a joint session of Congress, May 25, 1961 (David Whalen from George Washington University).

"Hazards of Communication Satellites," by J. R. Pierce, *The Bulletin of the Atomic Scientist,* May/June 1961.

"The systematic development of satellite communication systems," by K. G. McKay for presentation to the American Rocket Society, October 1961 (box 840902, AT&T archives).

"The Commercial Uses of Communications Satellites," by Leland S. Johnson (The RAND Corporation, June 1962).

"Aeronautical and Astronautical Events of 1961," report of NASA to the Committee of Science and Aeronautics, U.S. House of Representatives, June 7, 1962.

"Communication by Satellite," by Leonard Jaffe, *International Science and Technology,* August 1962.

"The dawn of satellite communication: a cooperative achievement of technology and public policy," by John A. Johnson (HAC archives).

"Communication satellites," *Journal of Spacecraft and Rockets* 14 (7), July 1977 pp. 385–394.

"Benefits in Space for Developing Countries," by Theo Pirard, *Aerospace International* May/June 1980.

"Satellite links get down to business," *High Technology Magazine,* June 1980.

"Rocky Road to Communication Satellites," draft of material prepared by Barry Miller for a lecture in the early 1980s (HAC archives).

"The History and Future of Commercial Satellite Communication," by Wilbur L. Pritchard, *IEEE Communications Magazine 22* (5), May 1984.

"The Bell System," *Encyclopedia of Telecommunications* (Marcel Dekker, 1991).

"The American Telephone and Telegraph Company (AT&T)," *Encyclopedia of Telecommunications* (Marcel Dekker, 1991).

"History of Engineering and Science in the Bell System," in *Transmission Technology,* edited by E. F. O'Neill (AT&T archives 85–70382).

"The Development and Commercialization of Communication Satellite Technology by the United States," by George Hazelrigg Jr. Draft in the NASA History Office.

How the World was One, by Arthur C. Clarke (Bantam Books, 1992).

Index

A

Active broadband satellite
  communication, 186, 191
Adams, Charles Francis, 209
Advanced Research Projects Agency
  (ARPA), 99
  Discoverer satellites of, 107
  and TIROS, 138, 145
  twenty-four-hour satellite
    committee of, 185
  and weather satellites, 142
Advent, 177, 215, 217–218
Air Force, 140
  Defense Meteorological Satellite
    Program (DMSP), 131, 147
  and history of satellite meteorology,
    131
  and Hughes Aircraft Company's
    proposal, 214
  interest in weather satellites, 138, 141
  involvement in tracking *Sputnik I,* 41
  and missile development, 25
  and TIROS, 146
  and Titan ICBM, 29
Albedo, 152–153
Allegany Ballistics Laboratory, 56
Allis Chalmers, contracts with Air
  Force, 141
Amateur astronomers. *See*
  Moonwatchers
American Meteorological Society, 140
  Suomi's lecture at, 168
American Museum of Natural History,
  31
American Rocket Society, 124, 214.
  See also *Jet Propulsion*
Antonov, Oleg, 16

Application Technology Satellite–I.
  See *ATS–1*
Applied Physics Laboratory (APL), 49
  and contamination of spacecraft, 102
  observations of satellite motion, 110,
    111
  and Polaris submarines, 51
  and TRAAC, 114–115
  and Transit, 56–57, 93, 97–99
  and Univac 1103A, 74
Argument of perigee, 61
Army, and Hughes Aircraft Company's
    proposal, 214. *See also* Jet
    Propulsion Laboratory (JPL);
    Project Orbiter
Army Air Forces, 141
Army Ballistic Missile Agency
    (ABMA), 121, 145
  and V2 rockets, 22
Army Signal Corps, Physics Research
    Laboratory of, 103
Ascending node, 61
ATS–1, 166, 167
  launch of, 168
AT&T, 103
  and Comsat, 179
  and *Echo,* 187
  monopolistic status of, 178–179
  plan for medium-altitude satellites,
    212
  proposal for global communication
    satellite system, 192–193
  relations with NASA, 178, 187,
    193–194, 197
  and *Telstar* (*see* Project Telstar; *Telstar*)
  transatlantic cables of, 171, 175–176
  See also Bell Telephone Laboratories